THE WORLD ON PAPER

ITEM 22. OLIVA. PORTOLAN CHART OF THE WESTERN MEDITERRANEAN. MANUSCRIPT. CA. 1590. 9 1/2" X 11 1/16"

THE WORLD ON PAPER

A Celebration of the Mapmaker's Art

BY MEAD T. CAIN

CATALOGUE OF AN EXHIBITION · MARCH 3 – JUNE 3, 1994

KEMPNER EXHIBITION ROOM

RARE BOOK & MANUSCRIPT LIBRARY

NEW YORK

COLUMBIA UNIVERSITY LIBRARIES

1994

ISBN 0-9607862-7-9

TABLE OF CONTENTS

ILLUSTRATIONS

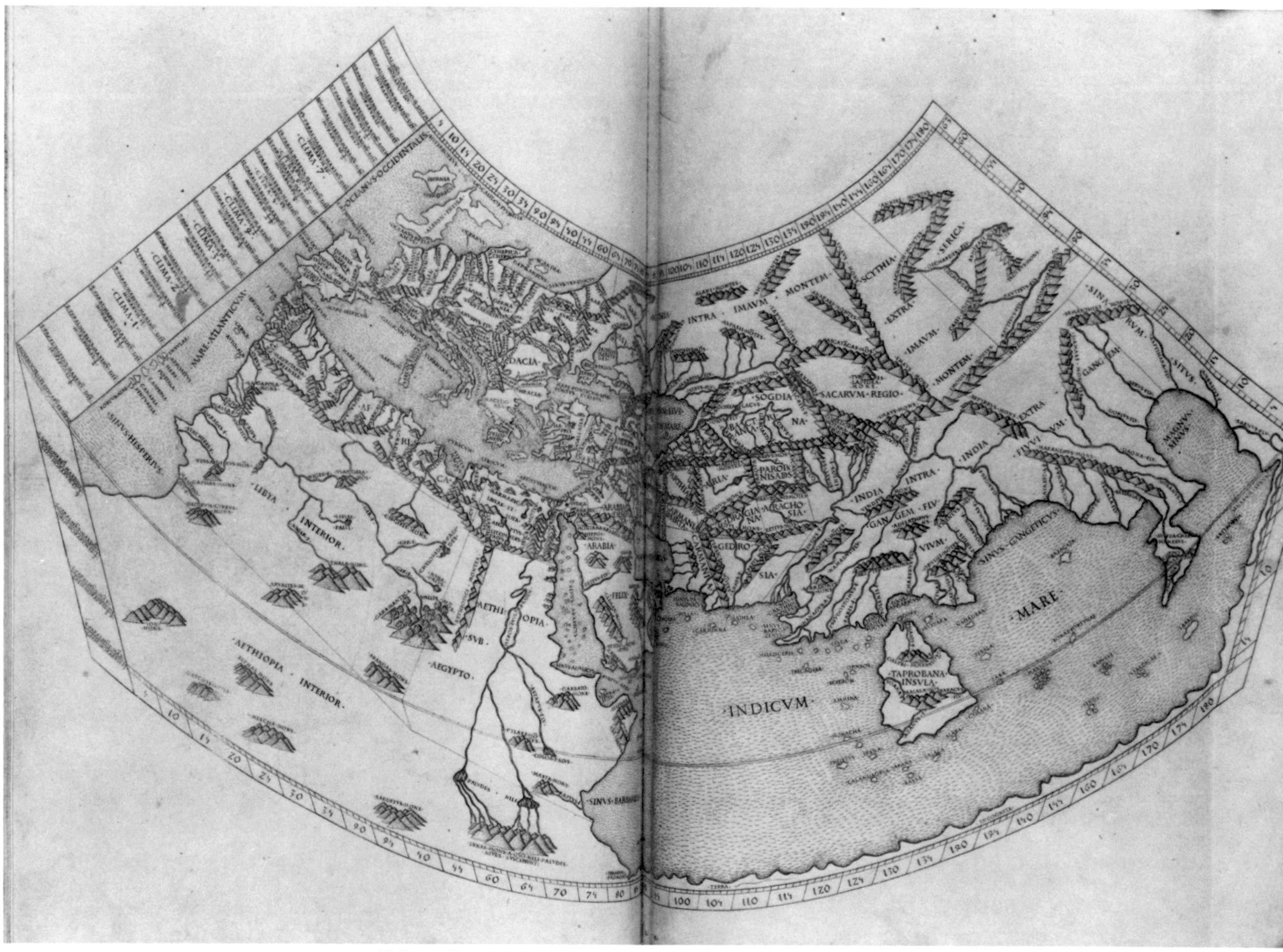

ITEM 7. PTOLEMY. WORLD MAP. 1490. 12⅝" X 22"

PREFACE

THE *World on Paper: A Celebration of the Mapmaker's Art* had its genesis in necessity. With a new director of the Rare Book and Manuscript Library still learning about the contents of the collections and the position of exhibitions librarian temporarily vacant, the postponement of a scheduled exhibition threatened to cause chaos and exhaust the resources of the staff. "Why not look at the map collection?" someone suggested. Maps are colorful, popular, informative, and—the salient point—would be easy to mount on the rear walls of the Kempner gallery cases. Moreover, they had not been on display for many years and thus were largely unknown to the Columbia community and the scholarly public. We would be able to provide access to materials that had been nearly hidden. Greatly aided by a 1931 published catalogue, *The Map Room of Columbia University,* and by entries in the local card catalogue, we quickly became aware that our collection indeed contained treasures and rarities. A serious map exhibition was long overdue. Our problem seemed to be solved.

We were lucky to find Mead Cain to help. An experienced map collector and scholar, Mead, as guest curator, brought to us a considerable knowledge of historical cartography and a lively interest in the exhibition process. Excited by some of the well-preserved examples of early maps in the Rare Book and Manuscript Library collections, Mead during his research quickly uncovered our even more significant collection of early atlases—many of them containing the same maps in their original context or, equally important, bound together into composite atlases that often captured the sensibility and geographical sophistication of the period of their construction. It soon became clear that we were no longer talking about an "easy" exhibition; despite the numerous problems involved with the display of large, fragile books and codex manuscripts, we were persuaded that inclusion of these volumes would significantly enrich the value of the exhibition to students and scholars. Jane Rodgers Siegel, Rudolph Ellenbogen, and eventually the newly appointed exhibitions librarian, Claudia Funke, worked closely with the guest curator to solve the numerous difficulties of installation and design. The final product, represented in part by the contents of this catalogue, was well worth the effort.

The maps and atlases in the Rare Book and Manuscript Library are the result of years of generous donations by alumni and Friends of the Library, among them most notably, George Arthur Plimpton, whose 1936 gift of medieval and Renaissance manuscripts and later texts has been part of so many enterprises undertaken by the Library. The current catalogue is made possible by further and more recent generosity: we are particularly grateful for the gift from Swann Galleries, which came through its president, George Lowry, and made the idea of publication a reality. We also thank Frank Streeter, Herbert Robinson, Charles Tanenbaum, Martayan Lan Rare Maps, and donors to the Rare Book and Manuscript Library Fund.

Maps and atlases provide a point of intersection between the physical world, the perceiving human eye, the imperial impulse, and the precise, sometimes exuberant skills of the cartographer. We hope *The World on Paper: A Celebration of the Mapmaker's Art* will do its part to inspire a new generation to look with curiosity and excitement at these magnificent records of the past.

JEAN ASHTON
Director, Rare Book and Manuscript Library

INTRODUCTION

IN 1894, shortly before his death, Robert Louis Stevenson wrote an essay on the origins of *Treasure Island* for a collection of reminiscences by famous authors about their first books. He told of sharing a house in Scotland one rainy day with a young boy "home for the holidays" who passed the time with the aid of "pen and ink and a shilling box of water-colors." Stevenson assisted the boy primarily by hanging his work, but sometimes he joined the artist at the easel:

> On one of these occasions I made the map of an island; it was elaborately and (I thought) beautifully colored; the shape of it took my fancy beyond expression; it contained harbors that pleased me like sonnets; and with the unconsciousness of the predestined, I ticketed my performance *Treasure Island.* I am told there are people who do not care for maps, and find it hard to believe. The names, the shapes of the woodlands, the courses of the roads and rivers, the prehistoric foot-steps of man still distinctly traceable up hill and down dale, the mills and the ruins, the ponds and the ferries, perhaps the *Standing Stone* or the *Druidic Circle* on the heath; here is an inexhaustible fund of interest for any man with eyes to see, or tuppenceworth of imagination to understand with.[1]

This exhibit is a celebration of early maps, as beautiful works of art and craft and as transports for our imaginations through time and space. It has become fashionable to eschew such attributes in favor of deeper meanings. Maps, like everything else, have become texts to deconstruct. Accordingly, one is invited to appreciate the power that maps wield, the deceptions and distortions that they contain, the subtle and sometimes sinister ways in which they manipulate one's consciousness and unconscious.[2] There is much to be said for this point of view; but it should never be permitted to detract from the pleasure that maps can give any man or woman "with eyes to see or tuppenceworth of imagination." (It is refreshing to remember John Aubrey's description of the eminently serious Thomas Hobbes, who "when at Oxford . . . used to love to visit the bookseller's shops, there to lye gaping on maps."[3])

Although the map enthusiast needs no excuse for celebrating the mapmaker's art, it should be noted in the case of this exhibit that an additional reason for selecting such a diffuse theme was expedience: a block of time and exhibition space opened up unexpectedly, requiring that we settle on an exhibition title before a full assessment could be made of the scope and depth of the Rare Book and Manuscript Library's collection. Much like a generic seminar title given to an organizer long before the event, the title we opted for gave us as much latitude as possible in making final choices. In fact, it was clear from the ensuing assessment that Columbia's collection of early maps and atlases is exceptionally fine and could (and perhaps in the future will) support exhibitions of comparable size with a variety of more focused themes, ranging from Civil War mapping, through a history of urban design, to the story of the great national and colonial trigonometric surveys of the eighteenth and nineteenth centuries. These examples indicate only a few of the strengths of the collection.

The present exhibition presents a broad sample of the cartographic materials in the Rare Book and Manuscript Library. The exhibit—which brings together seventy-eight maps, atlases, books, and letters—is organized in rough, chronological order, beginning with ancient and medieval geographical conceptions and the earliest printed maps, and proceeding through the age of discovery to the decorative heights of the late-sixteenth and seventeenth-century Flemish and Dutch masters. National and thematic criteria then take over, with maps and atlases from France, England, and Germany, followed by a selection of material relating to the settlement of North America, sea charts, and, finally, celestial maps.

The earliest item in the exhibition is a twelfth-century schematic map of the world, drawn on vellum, from a French manuscript copy of Saint Isidore of Seville's *Etymologiae* (item 4); the most recent is a section of the large-scale map of France published in 1803—a copperplate engraving on paper—based on the first trigonometric survey ever conducted on a national level (item 45). The latter is one of five items in the exhibit relating to the Cassini family, which, over four generations, took a lead in establishing France as a center of "scientific cartography." Other notable selections include: a fine copy of the 1490 Rome edition of Ptolemy's *Geographia* (item 7); the rare German edition of Joan Blaeu's *Atlas Major* (items 30–33); one of two hundred impressions made in 1768 of the engraving showing the Mason-Dixon line, in its original, uncut form (item 65); and two ink and wash maps of the upper Mississippi river, from the manuscript Jay Treaty letterbook, executed by John Trumbull, who acted as John Jay's secretary during the 1794 treaty negotiations (item 66).

An effort has been made in the commentaries to maintain a semblance of narrative coherence, so that the reader can take away from them a sense of how the artifacts relate to one another, their historical and scientific significance, and a crude outline of the development of cartography as an art and science in western Europe—the source of all but two of the items in the exhibit. The resulting outline

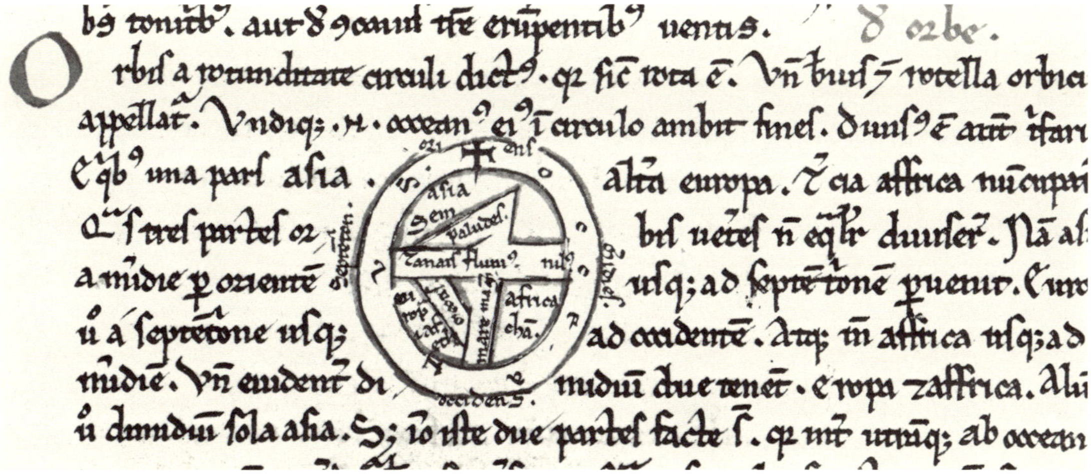

ITEM 4. ISIDORE OF SEVILLE. T-O WORLD MAP. MANUSCRIPT. LATE 12TH CENTURY. 1 1/16" DIAMETER

has obvious deficiencies and no claim is made for completeness. Any success in achieving the desired narrative coherence in an enterprise such as this is a function of the richness of both the collection and the reference sources at one's disposal. With respect to the latter, it is appropriate here to acknowledge several important intellectual debts. The first is owed to the organizers of *The World Encompassed*, an exhibition of the history of maps held at the Baltimore Museum of Art in 1952, which assembled 288 objects from all over the world. The catalogue from this most comprehensive exhibit provided an initial guide to my survey of the Rare Book and Manuscript Library's collection, and its influence can be seen in the organization of the present exhibition. Two items in our exhibit were borrowed for display in *The World Encompassed*: the twelfth-century manuscript *Etymologiae* of Isidore referred to above and a magnificent portolan atlas of five sea charts by Joan Oliva (item 22). I want also to mention two reference works that I found particularly helpful in compiling this catalogue: Tony Campbell's *The Earliest Printed Maps: 1472–1500* and Robert Karrow's *Mapmakers of the Sixteenth Century and Their Maps.*

It remains for me to express my gratitude to all those who helped make both the exhibition and this catalogue possible. Claudia Funke mounted the exhibition and provided important editorial, bibliographic, and conservation guidance. Rudolph Ellenbogen applied his editorial skills to the process, fashioning elegant copy from often verbose prose and supervising the production of the catalogue. Jane Siegal helped mount the exhibit and contributed in many other ways. Consuelo Dutschke provided timely and authoritative translations and bibliographic advice. Kevin

O'Connor helped with typing and formatting of the entries. Mary Pedley kindly supplied information on J. B. Nolin's map of Jesuit activities in France. Tony Campbell read an earlier version of the catalogue entries for errors. The author is, of course, responsible for all remaining errors. I thank Jean Ashton, the director of the Rare Book and Manuscript Library, for suggesting this exhibition in the first place, and thus providing me with such a wonderful opportunity and experience. She has been most supportive throughout and also contributed substantially to the editorial process. It is through her inspired fundraising efforts that this catalogue was made possible. Finally, I thank my wife, Gabrielle, and daughter, Lillian, for their moral and intellectual support over the course of this project, which, certainly for the month prior to the opening of the exhibit, absorbed virtually all of my attention.

MEAD T. CAIN
Guest Curator

1. Robert Louis Stevenson, "My First Book: *Treasure Island*," reprinted in *The Lantern-Bearers and Other Essays*, ed. Jeremy Treglown (London: Chatto & Windus, 1988), p. 279.
2. See, for example, Denis Wood, *The Power of Maps* (New York: Guilford, 1992).
3. Quoted in R. A. Skelton, *Decorative Printed Maps of the Fifteenth to the Eighteenth Centuries* (London: Staples Press, 1952), p. 1.

THE WORLD ON PAPER

A Celebration of the Mapmaker's Art

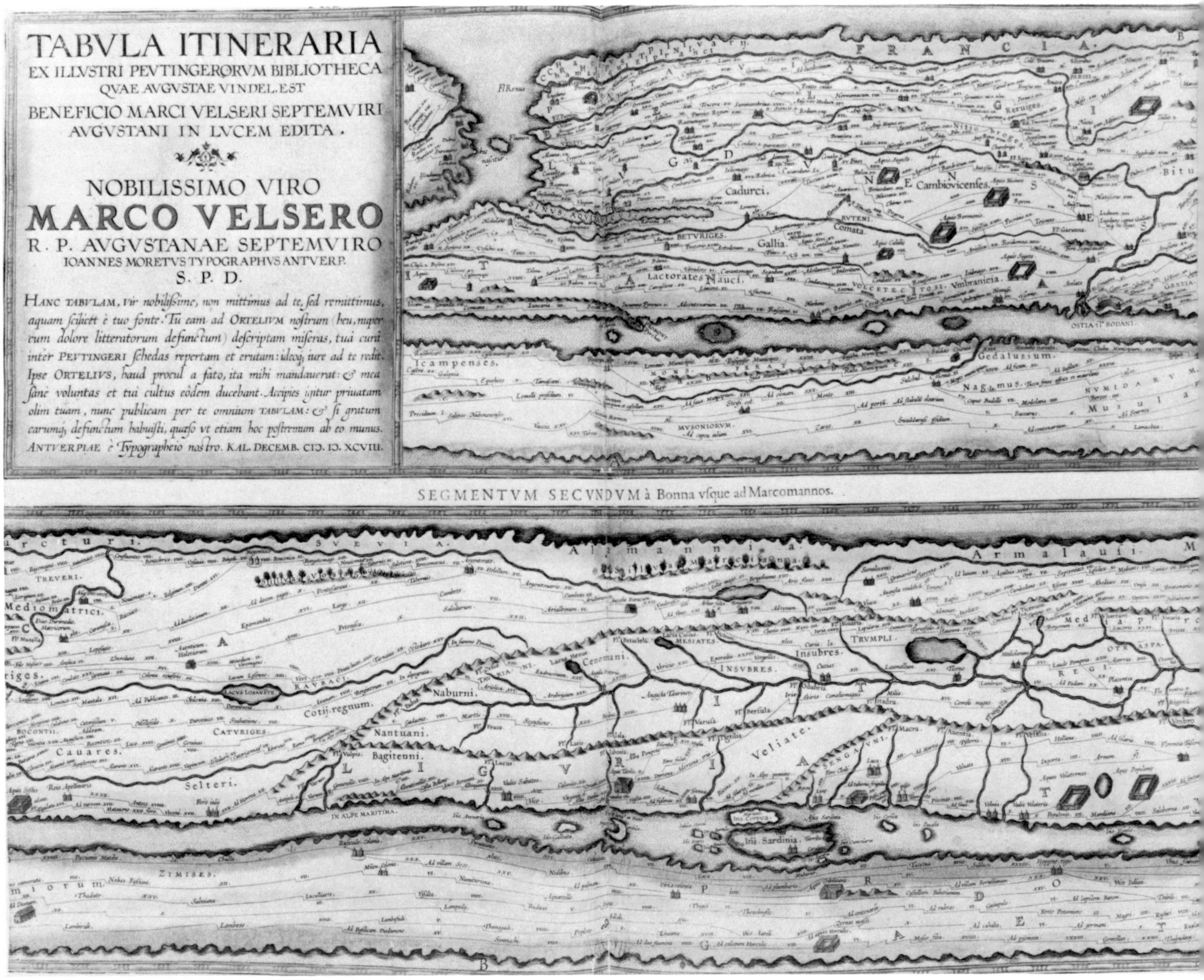

ITEM I. 7 5/8" X 20 3/8" EACH SEGMENT

I. MAPS OF ANTIQUITY & THE MIDDLE AGES

1. Peutinger Table

Segments I & II

In Petrus Bertius, *Theatrum Geographiae Veteris*

Leyden and Amsterdam: Isaac Elsevier for Jodocus Hondius, 1618–1619

The map known as the Peutinger Table is one of the few important cartographical remains of the Roman Empire. It is a map of the military roads in the Roman Empire in the fourth century A. D. The original is lost, but a copy was made in the twelfth or early thirteenth century and is now preserved in the Nationalbibliothek, Vienna. First discovered by the German humanist Konrad Celtes (1459–1508), it was willed by him to Konrad Peutinger (1465–1547). The entire map, consisting of the eleven extant sections, measures 22 feet 1 5/8 inches long and 13 3/8 inches high. Two sections of the Peutinger Table were first published in a much-reduced facsimile along with commentary in 1591. In 1598, eight of the eleven sections were reproduced in copperplate engraving by Abraham Ortelius. The present example shows two segments from the Ortelius imprint, which appeared in quantity for the first time in 1619. The map was not reproduced in its entirety until 1753.

Koeman, 1967, vol. I, Ber 1; *World Encompassed*, no. 5; Karrow, p. 27; Dilke, 1987, p. 238; Shirley, no. 212; Dilke, 1985, p. 113.

2. Ambrosius Aurelius Theodosius Macrobius, ca. 395–436

In somnium Scipionis expositio

Venice: Joannes Rubeus Vercellensis, 1492

In about 430 A. D. Macrobius, the late Roman philosopher and grammarian, wrote a commentary on "The Dream of Scipio" (Book VI of Cicero's *De republica*) that was illustrated by a world map. Macrobius described a spherical world with four main land masses, each surrounded by oceans. His map shows only the Eastern Hemisphere, with the known world crammed into the northern half above a vast and impassible equatorial sea. Overlaying the map are the five climatic zones of classical Greek cosmography. The two polar zones were thought to be frigid and uninhabitable and the equatorial zone impassible because of its heat. Only the two temperate zones between these extremes were habitable.

The world map in this edition of Macrobius is a crude and distorted version of the first printed edition of the map, produced in Brescia in 1483: Babylon and Jerusalem have been transposed on it; India has been disconnected from the mainland; and several of the smaller place names—including "Britani" and "India"—have been reversed in the cutting.

Campbell, 1987, p. 114; Shirley, no. 13; *World Encompassed*, nos. 7, 8; Woodward, 1987, p. 300; Suarez, no. 3.

Bequest of Gonzalez Lodge

3. Johannes Eschuid, d. 1380

Summa astrologiae judicialis

Venice: Johannes Lucilius Santritter for Franciscus Bolanus, 1489

Another copy of the 1483 Macrobius map appeared in the 1489 edition of *Summa astrologiae* by Johannes Eschuid, an English physician and astrologer. In this version the geographical representations have been reversed in the cutting, so that Britain is now to the northeast and India and Taprobana are hard against the western edge of the hemisphere. India has again been severed from the rest of the continent; however, this is otherwise a faithful copy and far superior to the 1492 effort.

Campbell, 1987, p. 115; Shirley, no. 13.

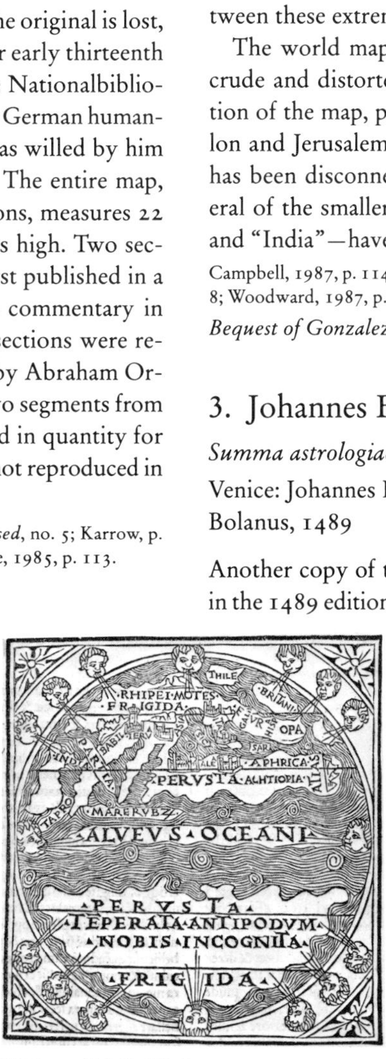

ITEM 3. 3 7/8" x 3 11/16"

4. Saint Isidore of Seville, ca. 560–636

Etymologiae

Manuscript on vellum

France, late 12th century

Rivaling Macrobius's zonal world map as a prototypical mappamundi for the Middle Ages was the T-O, or tripartite map, of the encyclopedist Isidore of Seville. Written between 622 and 633, his *Etymologiae* distilled the knowledge of the time. His geography and cosmography relied on both Roman and early Christian authorities. Maps, varying greatly in complexity, survive in Isidorian manuscripts from about 700 A.D. onwards. The map in this twelfth-century manuscript is relatively simple. Arranged with east at the top, the T within the O divides the world into the three continents of Asia (occupying the upper half of the circle), Europe, and Africa. The names of the continents are inscribed and paired with the sons of Noah who were reported to have settled each: Shem (Sem), Asia; Ham (Cham), Africa; and Japhet (Iafeth), Europe. The Don (Tanais) and Nile (Nilus) rivers form the cross of the T, and the Mediterranean (Mare Magnum), the stem. The V-shape on its side, to the left, is the Sea of Azov (Meotis Paludes). Around the rim of the world is an ocean (Occeanus [sic]). [ILLUSTRATION PAGE 11]

Campbell, 1987, pp. 1–2, 108; Brown, pp. 96–97; Woodward, 1987, pp. 299–301; *World Encompassed*, no. 9.

Gift of George A. Plimpton

5. Saint Isidore of Seville, ca. 560–636

Etymologiae

Augsburg: Günther Zainer, 1472

There were seven incunable editions of the *Etymologiae*, each with a map printed from a different woodblock. The map in this, the first edition, is one of the most rudimentary of the T-O variety, yet it has the distinction of being the first reliably dated map to appear in print. Less elaborate than the preceding manuscript T-O, it is without a separate Sea of Azov, for example, and neither the Don nor the Nile is mentioned by name. Although the original seventh-century Isidorian T-O has not survived, one can conjecture that it would have been a simple version like this. Recent scholarship has emphasized that it is inappropriate to infer anything about the evolution of contemporary geographic knowledge (or lack thereof) from these diagrams, because accurate representation of geographical form was not their primary function. They are better viewed as tools of moral education, employing "mystical, symbolic and spiritual imagery." Technical and cost constraints also argued for simplicity in design.

Campbell, 1987, pp. 2, 108; Woodward, 1987, pp. 301, 286–370 *passim*; Woodward, 1985; *World Encompassed*, no. 11; Shirley, no. 1.

American Type Founders Company Library

6. Werner Rolewinck, 1425–1502

Fasciculus temporum

Venice: Erhard Ratdolt, 1484

Rolewinck's *Fasciculus temporum*, originally published in 1474, was the first popular summary of history. The map appears in only two of the thirty-three known incunable editions, both printed by Erhard Ratdolt, the first in 1480, and the second in 1484. The map is of the T-O design, with the unusual feature of a large symbol for Jerusalem at the eastern end of the Mediterranean.

Campbell, 1987, p. 142.

II. THE REVIVAL OF PTOLEMY

7. Ptolemy, ca. 90–168

Geographia. Translated by Jacobus Angelus, with an index and treatise of Nicolaus Germanus

Rome: Petrus de Turre, 1490

Ptolemy, the Greek geographer, mathematician, and astronomer, composed his "Guide to Geography," known to the Renaissance as the *Geographia* or *Cosmographia*, in about 160 A.D. Although based on the learning of many predecessors, his work was the first attempt to systematize geographical knowledge and to place cartography on a scientific footing. In his text Ptolemy laid out the geographic coordinates of approximately eight thousand places, and the earliest surviving manuscript maps (from the late thirteenth century) conform to these specifications.

While the Middle Ages neglected Ptolemy in favor of Pliny, Pomponius Mela, and Solinus, he enjoyed a remarkable revival during the Renaissance. In the 1460s, Nicolaus Germanus redrew the twenty-six regional maps of the so-called "A" recension (one of two manuscript traditions) on a trapezoidal projection that served as a model for the 1478 Rome edition, printed by Arnoldus Buckinck.

The maps in the 1490 edition, displayed here, are from the unaltered 1478 plates. Although there was one nominally earlier printed edition containing maps (Bologna, 1477), work on the Rome plates started first (and even may have been completed first). As such these are the first maps produced by the process of copperplate engraving. Credit for the maps goes to Conrad Sweynheym, who worked on them from 1474 until his death in 1477. They are beautifully rendered and are often held to be the finest of all the many editions of Ptolemy's maps, rivalled only by Gerard Mercator's 1578 edition. Part of their aesthetic appeal is the elegant lettering, which has been punched from a set of dies, rather than cut by hand.

The world map, unlike the twenty-six regional maps, is drawn on a conic projection. The chief difficulty with this projection is the distortion created south of the equator. The solution was to "break" the map and to delineate those parts south of the equator by a corresponding set of south polar conic meridians.

[ILLUSTRATION PAGE 6]

Campbell, 1987, pp. 4, 122–127, 131–132; *World Encompassed*, nos. 4, 32, 33, 36, 40; Dilke, 1987, pp. 266–272, 276–279; Dilke, 1985, pp. 154–164.

American Type Founders Company Library

8. Pomponius Mela, fl. 37–42

Cosmographia, sive de situ orbis

Venice: Erhard Ratdolt, 1482

This unsigned map of the world is modeled closely on Ptolemy's. It is in most respects an unsophisticated copy, yet it incorporates important improvements in the representation of Africa's west coast. For the first time on a printed map it shows the southeastward turn in the coast south of the 12° parallel and the beginnings of the Gulf of Guinea. None of the incunable editions of Ptolemy reflected these discoveries.

Pomponius Mela is considered to be the earliest Roman geographer. The map bears little relation to Mela's geographical views, which included a belief in an inhabited southern temperate zone and would probably have been better served by a zonal mappamundi. While the compiler of this map is not known, it has been suggested that he may have been the printer Erhard Ratdolt, who is known for his technical innovations in printing and who might also have been drawn to geographical innovations. This map and Ratdolt's T-O illustration for the 1480 edition of Rolewinck were the first woodcut maps to be produced in Italy.

The inscription above the map has been translated as follows: "If, in a new lease of life, a man seeks to attain the wormlike paths of geography, he is bound to find the flowers that belong there for he deserves them."

Campbell, 1987, pp. 118–119; *World Encompassed*, no. 41; Dilke, 1987, p. 242; Shirley, no. 8.

Gift of George A. Plimpton

9. Hartmann Schedel, 1440–1514

Liber chronicarum

Nuremberg: Anton Koberger, 1493

Two woodcut maps appeared in the *Nuremberg Chronicle* of 1493, a Ptolemaic world map (shown here) and a modern map of north and central Europe. The world map is copied from the Venetian woodcut that was bound into Mela's *Cosmographia* of 1482. What makes the map stand out (in addition to its superior craftsmanship and aesthetic appeal) are several medieval flourishes: the sons of Noah—Shem, Ham, and Japhet—who hold the world in their hands, and the procession of monsters that decorates the space to the left of the map. The placement of the map in the *Chronicle* at the beginning of the "Second Age," which starts with the story of Noah, helps to explain his sons' presence on the map. The fanciful creatures to the left are printed from a separate woodblock and are derived from the immensely popular *Polyhistor* of Solinus.

Campbell, 1987, p. 152; Shirley, no. 19; Rücker; Brown, p. 86; *World Encompassed*, no. 44; Nordenskjöld, p. 39.

Gift of Acton Griscom

10. Ptolemy, ca. 90–168

Claudii Ptholemaei Alexandrini Liber Geographiae cum Tabulis et Universali Figura et cum Additione Locorum Quae a Recentioribus Reperta Sunt Diligenti Cura Emendatus et Impressus

Venice: Jacopo Pencio, 1511

Even before the first printed edition of the *Geographia* appeared in 1477, there was a tension between the esteem in which Ptolemy was held by Renaissance scholarship and the rapid pace of discovery that invited a rewriting of his geography. In the Venice edition of 1511, Bernardus Sylvanus Ebolensis set about amending Ptolemy's "foundation" maps to take account of modern discoveries. But the result was less than satisfactory—many of his alterations were arbitrary and ill-founded, and his incorporation of current knowledge was incomplete. In the untitled world map shown here, modifications are evident in the shape of Africa and the British Isles. Also, the strip of land to the south that made the Indian Ocean an enclosed sea in the Rome editions of Ptolemy has been removed. The awkwardness of his revision is nowhere more apparent than in his treatment of India,

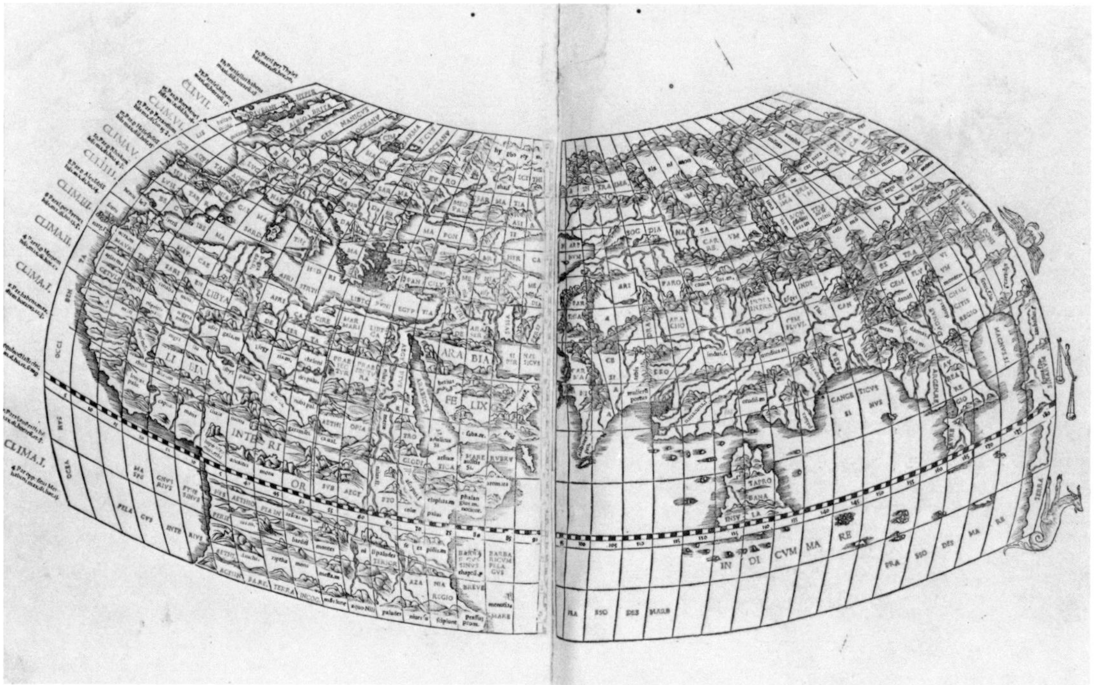

ITEM 10. 11½" x 22"

which has a "modern" peninsula grafted onto the Asian continent to the west of the traditional stubby Ptolemaic subcontinent. This does not lessen the visual appeal of the map, printed in two colors for the first time. Lower case black lettering is reserved for lesser geographical features, while red capital letters are used for important landmarks.

Shirley, no. 31; *World Encompassed*, no. 55; Nordenskjöld, p. 18; Campbell, 1987, p. 4; Brown, p. 74.

Gift of Dr. Dallas Pratt

11. Gregor Reisch, d. 1525

Margarita Philosophica

Strasbourg: Schott, 1504

This untitled Ptolemaic world map is found in a number of early editions of Gregor Reisch's *Margarita Philosophica*. It is unremarkable, with the exception of the twelve wind-heads—one bespectacled—which have considerably more personality than the cherubs that usually perform this task, and a Latin inscription in the strip of land bordering the Indian Ocean to the south: "Here is not land but ocean. In it are islands of extraordinary size unknown to Ptolemy." This has been interpreted by some as a reference to Columbus's discoveries, and if true, would constitute the first such reference on a printed map. The text of this edition of *Margarita Philosophica* betrays no awareness of these or other discoveries of the fifteenth century, and it is generally assumed that the map was inserted by Johann Schott, the publisher.

Nordenskjöld, p. 40; Shirley, nos. 22, 23; Suarez, no. 10; *World Encompassed*, no. 49; Grafton, p. 57.

Gift of Frederick A. P. Barnard

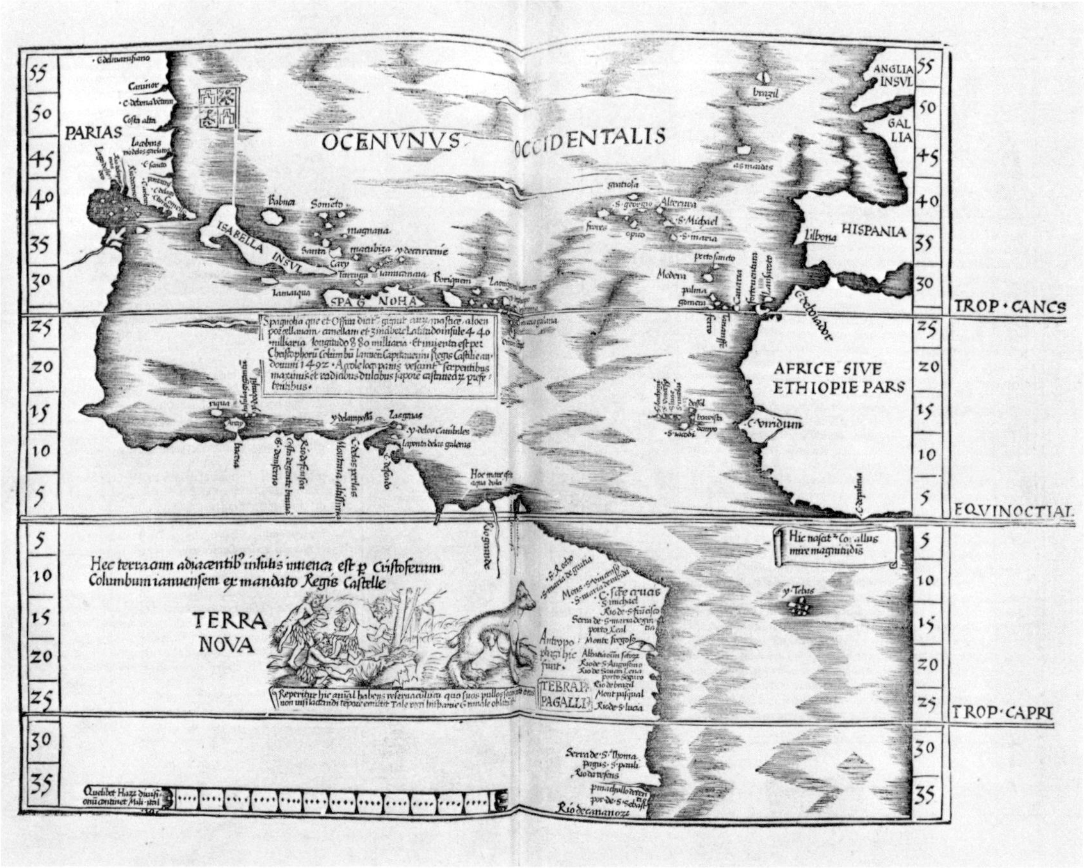

ITEM 12. 11¼" X 17"

III. MAPS FROM THE AGE OF DISCOVERY

12. Ptolemy, ca. 90–168

Geographia. Translated by Willibald Pirckheimer and edited by Michael Villanovano

Lyon: Melchior and Gaspar Trechsel, 1535

This 1535 edition of Ptolemy contains maps that copy those from the seminal 1513 edition of Martin Waldseemüller. Waldseemüller was highly critical of the earlier attempts to bring Ptolemy up to date by inserting modern maps or (as in the 1511 edition) correcting the old ones. He was the first to separate the modern from the ancient maps and to emphasize the historical nature of Ptolemy's maps. The map shown here, *Tabula Terrae Novae*, reproduces the first printed map devoted to the New World to appear in an atlas. It was Waldseemüller who, on his 1507 world map, first used the name "America," after Amerigo Vespucci, for the South American continent. Here he abandons the term and specifically credits Columbus with the discovery.

Calvin ordered the 1535 Lyon edition to be destroyed because of the alleged heresies that it contained, and thus it is rare.

Phillips, no. 364; *World Encompassed*, no. 56; Nordenskjöld, p. 22; Stevens, p. 48; Skelton, 1975, p. 66; Suarez, no. 16; Shirley, nos. 34, 48.

Gift of George A. Plimpton

13. Peter Apian, 1495–1552

Tipus Orbis Universalis

In Caius Julius Solinus, *Polyhistor*

Vienna: Johann Singriener, 1520

Waldseemüller's world map of 1507 is the model for this map by Peter Apian, published in 1520. It shares

ITEM 13. 11¼" X 16"

with the earlier map a similar geographical content, truncated cordiform projection, and woodcutting style. The style has been described as "robust," but Nordenskjöld, who possessed the bluntness of a seaman, declared that it was "badly drawn and badly engraved." A slim North America is shown separated by sea from South America. The retention of the name "America" on the southern continent and the geography of Asia, with almost no sign of the Portuguese discoveries, suggest that Apian did not consult Waldseemüller's later work. There is no reference to the map in the text, but the initials of the author of the commentary on Solinus, (Jo)hann (K)amers, appear in the lower left corner of the map. Apian was a mathematician, astronomer, geographer, and printer, but not a cartographer. His roles in map production seem to have been limited to those of editor and publisher. This was his first published work.

Karrow, pp. 49–63; Nordenskjöld, p. 99; Wolff no. 81; *World Encompassed*, no. 61; Suarez, no. 13; Shirley, no. 45.

Gift of George A. Plimpton

14. Johannes Honter, 1498–1549

Rudimenta Cosmographica

Zurich: Froschauer, 1548

Johannes Honter was born in Kronstadt (now in Romania) and entered the university in Vienna in 1515 at the same time as Peter Apian. In 1530, while at the university of Kraków, he published the first edition of what would become an exceptionally popular cosmography. He later returned to Kronstadt, where he established the first printing press and devoted himself to the spread of humanism. He concentrated on the production of school books, and for this purpose he transformed the text of his earlier cosmography into Latin verse (for ease in memorization by students) and created new maps. The first edition of the revised *Rudimenta Cosmographica* was published in Kronstadt in 1542. The maps are modern, not Ptolemaic, and one could claim that they constitute the first modern world atlas. Nearly ninety editions of *Rudimenta Cosmographica* printed between 1542 and 1692 have been identified, one third of which contain maps. Each publishing center had its own set of woodblocks, and the first recutting of the blocks occurred in Zurich in 1546. The displayed example is the second Zurich edition of 1548. Among the editions of Honter's maps, the Zurich impressions are unique in that place names are printed from letterpress type.

Karrow, pp. 302–315; *World Encompassed*, no. 66; Shirley, nos. 65, 79, 86.

Gift of George A. Plimpton

15. Manuscript Maps after Münster

Copied from Caius Julius Solinus's *Polyhistor*

Basel: Michael Isengrin and Henricus Petri, 1538

Ink and watercolors on paper, signed P. I. Pignon, 1542

In a *Sammelband* with Pomponius Mela, *De situ orbis libri III*

Paris: Jean Loys, 1541

This composite volume of texts, collected and bound together by its owner, contains mostly printings of classical works, including Mela's *De situ orbis libri III*. Seven manuscript maps, copied, probably by Pignon, from woodcuts in Sebastian Münster's edition of Solinus's *Polyhistor*, are bound in following the Mela text. The entire compilation, liberally annotated by the owner, mainly with transcriptions of commentaries by Adrianus Turnebus, provides a valuable glimpse of how such texts were used by a gentleman-scholar. [FRONT COVER ILLUSTRATION]

Karrow, p. 420.

16. Sebastian Münster, 1488–1552 Hans Holbein, 1497–1543

Typus Cosmographicus Universalis

In Simon Grynaeus and Johann Huttich, *Novus Orbis Regionum ac Insularum Veteribus Incognitarum*

Basel: Johann Herwagen, 1555

This map's authorship is now confidently attributed to Sebastian Münster, with the decoration on the map and its border only slightly less confidently attributed to Hans Holbein. It, too, is based on Waldseemüller's 1507 world map, or perhaps one of its derivatives, and was thus out of date by the time of its first publication in 1532. Whatever is lost in accuracy, however, is made up for in rich decoration. Two cherubs work cranks to turn the earth on its polar axis; vignettes representing the four continents fill the corners (with New World cannibals tucking into a meal in the lower left); and monsters, a mermaid, and a galleon enliven the sea. Münster was a prodigious and prolific scholar, being accomplished as a geographer, cartographer, Hebraist, and theologian. Aside from his cartographic and related output, which ran to over twenty books and pamphlets and almost two hundred separate map images, he published some eighty books on theological topics and Hebrew.

Karrow, pp. 410–434; Shirley, no. 67; Harley, pp. 85–86; Wolff, no. 86; Nordenskjöld, p. 105; *World Encompassed*, no. 65.

17. Sebastian Münster, 1488–1552

Asia Maior

From Caius Julius Solinus, *Polyhistor*

Basel: Michael Isengrin and Henricus Petri, 1543

A small portion of the western coast of North America appears as "Terra Incognita" in the upper right hand corner of this map. It is thought to be the earliest printed map to show this part of the North American continent. Münster believed that mastery and critique of ancient geography were a necessary foundation for his planned modern geography and cosmography. His first effort in this endeavor was an edition of Solinus's *Polyhistor* and Mela's *De situ orbis*, published in 1538, with new maps and commentary.

Karrow, p. 420; *World Encompassed*, no. 71; Grafton, p. 97.

Gift of Mr. & Mrs. Valerien Lada-Mocarski

18. Sebastian Münster, 1488–1552

Cosmographia

Basel: Henricus Petri, 1554

When Münster first published his *Cosmographia* in 1544, it had a mere 659 pages, with 520 woodcuts. It

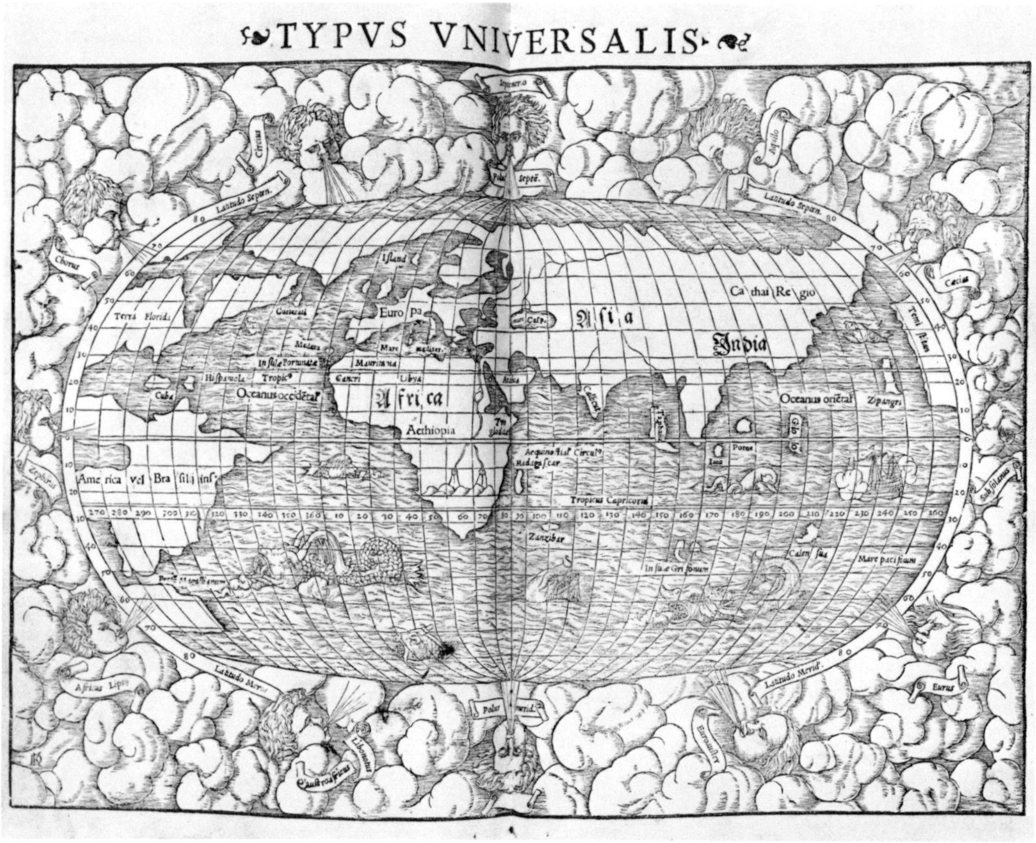

ITEM 18. 10 3/8" x 14 7/8"

grew steadily in successive editions to its present magisterial size (1,233 pages and 910 woodcuts), first achieved in the 1550 Basel edition and largely unchanged thereafter. Characterized as a "tidal wave of erudition," by one admirer, it surveyed all the lands and people of the world, describing in the process "everything from the dynasties of royal houses to the monsters that inhabit northern and eastern lands and waters." The *Cosmographia* was enormously influential, and it would thrive long after Münster died of the plague in 1552. Twenty-seven posthumous editions in German, French, Latin, Italian, and Bohemian, dated between 1553 and 1628, have been counted. The section on "newe India" was translated into English and published in 1553, making it one of the first three English books on America. According to one assessment, the *Cosmographia* "taught nearly three generations of laymen most of what they knew about the world beyond their native place."

This world map, from a block cut for the 1550 edition shows a marked improvement in crude geography over Münster's 1532 map.

Karrow, pp. 429–434; *World Encompassed*, no. 58; Shirley, no. 92; Grafton, p. 97; Ruland.

19. Sebastian Münster, 1448–1553

Tabula Novarum Insularum

In Münster, *Cosmographia*

Basel: Henricus Petri, 1550

This striking map of the New World, the first separate map of the continent as a whole, was created for Münster's 1540 edition of Ptolemy's *Geographia*. Among its interesting features are the scatter of islands in the Pacific ("Archipelagus 7448 insularum"), which originates in Marco Polo's writings, the island of Japan ("Zipangri") shown just off the coast of Mexico, and the deep cleft in the North American continent, which almost separates "Francisca" from "Terra florida." This cleft is the "oriental sea" reported by Verrazano, which according to him was separated from the Atlantic only by a long and narrow isthmus.

Wolff, no. 90; Harley, pp. 93–95; Suarez, no. 23; Karrow, pp. 421–434; Ruland; Nordenskjöld, pp. 108, 113.

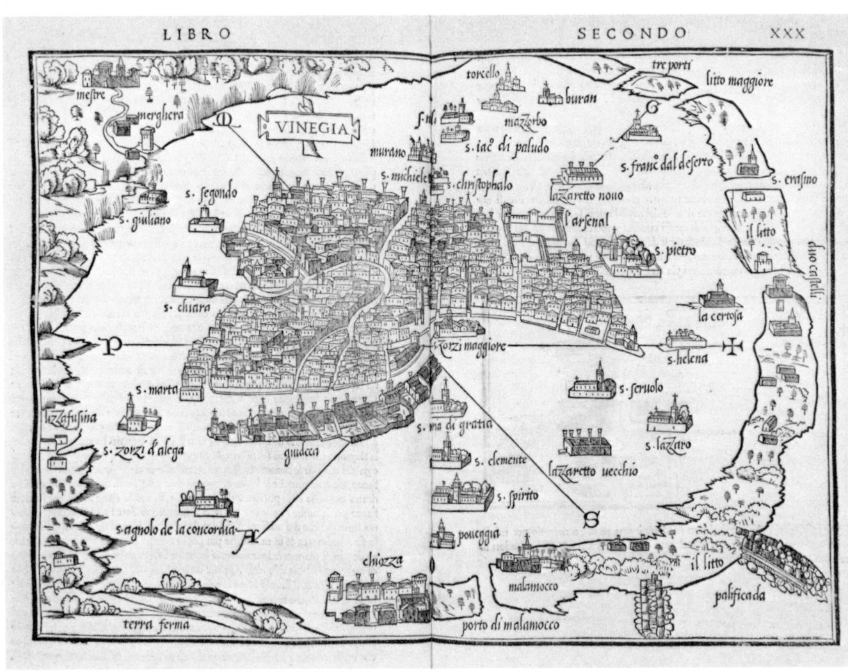

ITEM 20.
9" x $12\frac{1}{4}$"

20. Benedetto Bordone, d. 1539

Libro di Benedetto Bordone. Nel qual si ragiona de tutte l'isole del monde

Venice: Nicolo d'Aristotile, detto Zoppino, 1528

Isolari, or island books, probably had their origins in the late Middle Ages, along with manuscript portolan charts, as aids to navigation in the Mediterranean. The earliest known *isolario*, however—Cristoforo Buondelmonti's *Liber Insularum Archipelagi*, the model for Bordone's—was written for "a prince of the church and a scholar," by an "ecclesiastic with a taste for antiquities." The many manuscript copies of the *Liber* that survive from the fifteenth century attest to its demand in such circles. Bordone's popular work broke with earlier prototypes, which did not venture beyond the Mediterranean, by surveying islands worldwide.

Harvey, 1987, pp. 482–484; Karrow, pp. 89–93; Shirley, no. 59; Nordenskjöld, pp. 103–104; *World Encompassed*, no. 83.

Gift of George A. Plimpton

21. Tommaso Porcacchi, ca. 1530–1585

L'isole più famose del mondo, descritte da Tomaso Porcacchi da Castiglione Aretino

Venice: Pietr'Antonio Brigonci, 1686

The success of Bordone's project inspired this *isolario* by Porcacchi, first published in 1572. The maps are of much finer quality than Bordone's, both aesthetically and as geographical representations. They are copperplate engravings executed by Girolamo Porro. The enduring appeal of this little book is indicated by the date of its last edition—1713.

Hough, no. 100; *World Encompassed*, no. 86; Nordenskjöld, p. 104.

22. Joan Oliva, fl. 1580–1615

Portolan atlas of five charts of the European and African Coasts of the Mediterranean and Atlantic

Manuscript on vellum, signed, Iovanne Oliva fecit.

Italy, ca. 1590

The portolan chart is of the same tradition as the *isolario*, and many of the portolan atlases made by the Oliva family and other chart makers of the period include an *isolario* at the end. This fine example has only charts. Portolan charts were used by mariners well into the seventeenth century, but there was also a demand for richly decorated versions among the enlightened wealthy. One can assume that the present atlas was meant for this market. Joan Oliva was the most prolific member of a large family of Catalan chart makers, one branch of which had settled in Messina (Sicily) some time before 1550. Charts signed by at least sixteen members of the Oliva family are recorded, with dates between 1538 and 1673.

[FRONTISPIECE]

World Encompassed, no. 98; Howse and Sanderson, p. 45; Rey Pastor and García Camarero, p. 137.

Gift of George A. Plimpton

23. Ptolemy, ca. 90–168

La geografia di Claudio Ptolemeo . . . con alcuni comenti & aggiunte fattevi da Sebastiano Munstero . . . con le tavole non solamente antiche & moderne solite di stamparsi, ma altre nuove aggiuntevi di messer Iacopo Gastaldo . . . ridotta in volgare italiano da m. Pietro Andrea Mattiolo

Venice: Giovan Baptista Pedrezano, 1548

Giacomo Gastaldi, fl. 1539–1566, who made the maps for this edition of Ptolemy, was one of the most important cartographers of the sixteenth century and the greatest Italian mapmaker of his age. The period during which he was active was also a time when Italian—especially Venetian—engravers and publishers came to dominate the European map trade. Gastaldi's oeuvre, which included many separately published, large-scale maps of all parts of the world, would be much imitated and provided the models for many of the maps in the earliest modern Flemish world atlases published by Abraham Ortelius (1570) and Gerard de Jode (1578). If one were forced to mark the beginning of a golden age of cartography—whose highest expression was to be found in mid-seventeenth century Holland with the work of the Blaeu family—Gastaldi would be a good choice.

The sixty maps in this volume include twenty-six Ptolemaic maps copied from Münster and thirty-four modern maps, all of which are newly drawn and freshly conceived. The map shown—*Carta Marina Nova Tabula*—is one of two modern world maps. On it, North America is depicted as joined to both Asia and Europe.

World Encompassed, no. 122; Karrow, pp. 216–249; Shirley, no. 99; Suarez, no. 24; Wolff, no. 93; Nordenskjöld, pp. 25–26, 112–113.

Gift of George A. Plimpton

24. Girolamo Ruscelli, ca. 1504–1566

Orbis Descriptio

From Ptolemy, *La geographia di Claudio Tolomeo Alessandrino, Nuovamente tradotta di Greco in Italiano da Girolamo Ruscelli*

Venice: Vincenzo Valgrisi, 1561

Ruscelli's elegant, twin hemisphere world map is based loosely on an oval map of the world that appeared in Giacomo Gastaldi's 1548 edition of Ptolemy. It is not the first double hemisphere map (although it is one of the first), but it is the first such map to be found in an atlas. There is no "southern continent" (Antarctica); North America is joined to an arctic land mass; and Asia and North America are shown to be connected by a land bridge.

Shirley, no. 110; Karrow, pp. 222–223; Nordenskjöld, p. 26; Wolff, no. 94.

25. Giovanni Battista Ramusio, 1485–1557

Terzo volume delle navigationi et viaggi

Venice: Stamperia de' Giunti, 1565

The third volume of Ramusio's collection of discovery and exploration texts (first published in 1556) contains six woodcut maps by Gastaldi. The map *La Nuova Francia* accompanies the report of Verrazano's voyages. It shows the coast from New York Bay to Labrador. "Port du Refuge" is Narragansett Bay.

Gastaldi, who was a native of Piedmont, became closely associated with Ramusio soon after his arrival in Venice, sometime in the late 1530s. Ramusio was secretary of the senate at the time, and it was owing to his influence, perhaps, that Gastaldi first came to hold what would become a succession of official positions, including "cosmographer to the Republic of Venice."

Hough, no. 31; Karrow, pp. 216, 229; Suarez, no. 26.

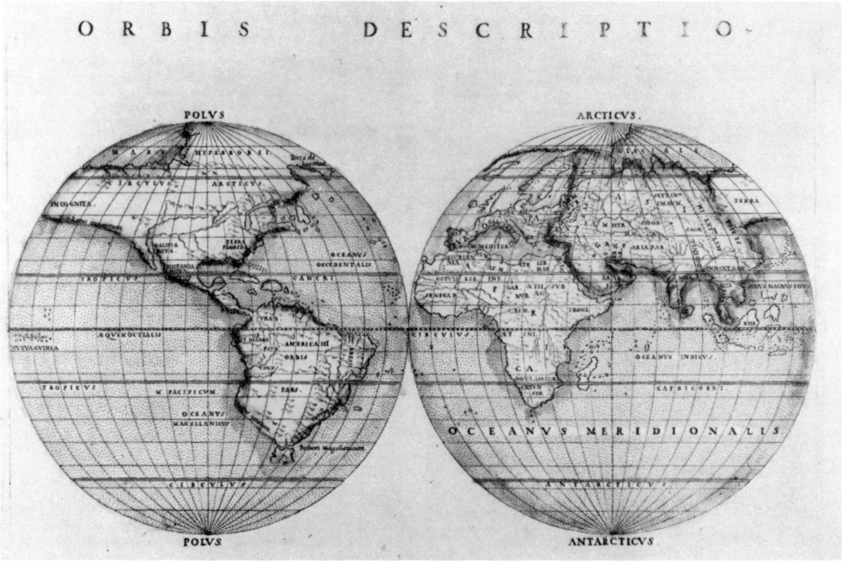

ITEM 24. 5" X

IV. THE GOLDEN AGE OF CARTOGRAPHY: THE LOW COUNTRIES

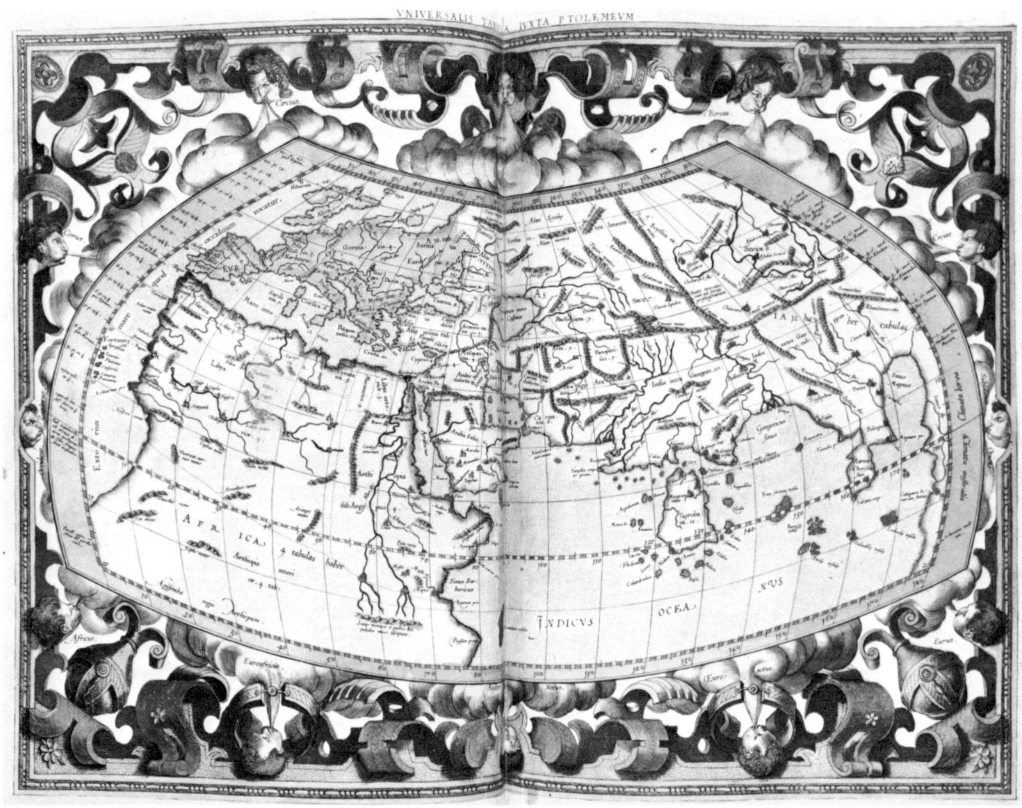

ITEM 26.
13 15/16" x 18 7/8"

26. Gerard Mercator, 1512–1594

Tabulae Geographicae

Cologne: Godefridus Kemp, 1578

Bound with:

Ptolemy, *Geographia*

Lyon: Melchior and Gaspar Trechsel, 1535

At the age of about fifty, Gerard Mercator, an accomplished cartographer and globe and instrument maker, conceived of an ambitious project—a massive cosmographical and geographical treatise that would comprehend the creation of the world, astronomy, ancient and modern geography, genealogical and political history, and chronology. The first part to be completed was the *Chronologia* in 1569. The next component of his grand project was this edition of Ptolemy's maps, beautifully engraved and showing a loving attention to detail in their adherence to the original specifications.

The final major component of Mercator's project, his modern geography, or *Atlas sive Cosmographicae Meditationes de Fabrica Mundi et Fabricati Figura,* would not be published in its entirety until 1595, a year after his death. The plates for the *Atlas* and the *Tabulae Geographicae* were sold in 1604 to Jodocus Hondius, of Amsterdam, and with his vigorous promotion, Mercator's work enjoyed great success over the next century. The use of the word "atlas" to signify a collection of maps originates with Mercator's title.

Koeman, 1967, vol. II, pp. 281–283; Karrow, pp. 376–406; *World Encompassed*, no. 133.

27. Abraham Ortelius, 1527–1598

Theatrum Orbis Terrarum

Antwerp: Plantin Press, 1598

Ortelius's *Theatrum Orbis Terrarum*, originally published with fifty-three maps in 1570, was the first modern geographical atlas. Unlike Mercator, who created his own maps, Ortelius was essentially a compiler: he gathered the best maps that he could find from all sources and had them re-engraved in uniform size. A unique feature of the *Theatrum* is a *catalogus auctorum*—a list of cartographers, and their maps, on whom Ortelius relied. Fifteen Latin editions appeared, the last in 1612; many additional editions were issued in other languages. The present example, the 1598 Dutch edition, is open to the map of Iceland, which is dominated by outlandish monsters that surround the island.

Koeman, 1967, Ort 31; Karrow, pp. 1–31; *World Encompassed*, no. 135.

28. Abraham Ortelius, 1527–1598

Americae sive Novi Orbis Nova Descriptio

From Abraham Ortelius, *Theatrum Orbis Terrarum*

Antwerp: Egidius Coppens Diesth, 1570

Ortelius's map of America, the second in the sequence of maps in the *Theatrum*, is especially attractive. It is reasonably accurate for its day, showing, for example, both the Baja and Yucatan Peninsulas. The strange "potato" shape of South America would be corrected in the second edition of this map, issued in 1587.

Karrow, no. 1/7.

29. Georg Braun, 1541–1622, and Franz Hogenberg, fl. 1558–1590

Civitates Orbis Terrarum, Liber Primus

Cologne, 1572

Bound with:

Ortelius, *Theatrum Orbis Terrarum*

Antwerp: Egidius Coppens Diesth, 1570

Braun and Hogenberg's *Civitates*, or townbooks, would eventually extend to six volumes and include almost four hundred plans, vistas, and bird's-eye views of cities and towns located throughout the world. The volumes were published serially over a forty-five-year period between 1572 and 1617. Like the *Theatrum* of Ortelius, *Civitates* was one of the most popular works of the last quarter of the sixteenth century. Although a product of Cologne, it belongs to the same cartographic school of the south-

ITEM 29. $13^{1}/_{4}$" x $18^{5}/_{8}$"

ern Netherlands that included the work of Mercator, Ortelius, and de Jode. The engraving reflects this style, and in all probability, Ortelius had a hand in planning the work. It is certain that he supplied much of the material for the *Civitates*. One can agree with Robert Burton who wrote in his *Anatomy of Melancholy*, "What greater pleasure can there now be than to view those elaborate Maps of Ortelius, Mercator, Hondius, &c. To peruse those books of cities, put out by Braunius and Hogenbergius."

In the volume on display, the title page, text, and engraved town plans from the first volume of *Civitates Orbis Terrarum* are bound together and interleaved with maps from the first edition of Abraham Ortelius's *Theatrum Orbis Terrarum* (Antwerp, 1570), so that the Braun and Hogenberg plans follow the Ortelius maps on which the towns appear. Shown here is the plan of Antwerp.

Koeman, 1967, vol. II, pp. 10–11; Skelton, 1952, p. 26; *World Encompassed*, no. 269.

30. Joan Blaeu, 1596–1673

Atlas Major, das ist, Welt-Beschreibung. Volume 1 of 9, *Nördliche Europa*

Amsterdam: 1667

Flemish and Dutch prominence in European cartography in the late sixteenth century would become dominance in the seventeenth century with the Amsterdam houses of Blaeu and Jansson. While both were very successful and produced comparable output, there is little question that the Blaeus were preeminent. Their crowning achievement, described as "the greatest and finest atlas ever published," was the *Atlas Major*, which appeared in 1662 in eleven volumes with 593 maps. Joan Blaeu, the son and main successor to Willem Janszoon Blaeu (1571–1638), who founded the firm, supervised the project. The maps comprising the atlas were long in the making, and they represent the collective output of two generations. The *Atlas Major* was issued in Latin, Dutch, French, Spanish, and German editions. The Columbia University copy is the exceptionally rare German edition, of which only one other complete example is known. Its rarity has prompted the suggestion that copies were assembled on commission for customers in Germany, and German title pages were printed for them. The nine-volume atlas includes the 605 maps called for in Koeman (*Atlantes Nederlandici*) as well as many additional maps from other cartographers, including Visscher, de Wit, Allard, Jansson, and Homann, among others.

Two world maps are supplied in this copy of *Atlas Major*, an earlier (1606) map by Willem Janszoon Blaeu with distinctive decorative border panels, and this beautifully colored map by Joan Blaeu (*Nova et Accuratissima Totius Terrarum Orbis Tabula*), which he made specially for the *Atlas Major*. Above the twin hemispheres of the world are celestial figures reclining on clouds. Two learned and somewhat glum-looking cosmographers are shown, one holding an armillary sphere and the other a set of dividers. Along the bottom are the four seasons, each of whom is seated on a chariot drawn by animals.

Koeman, 1967, vol. I, pp. 68–69, Bl 61, Bl 56 [435]; Brown, pp. 171–175; *World Encompassed*, no. 149; Shirley, no. 428.

31. Claes Janszoon Visscher, 1587–1652

Marchionatus Sacri Romani Imperii

In Joan Blaeu, *Atlas Major, das ist, Welt-Beschreibung*. Volume 3 of 9, *Niederlandt*

Amsterdam: 1667

Bound in the Columbia University copy of Blaeu's *Atlas Major* are a large number of maps by other cartographers, both contemporaneous with the atlas and from later periods. Given the presence of eighteenth-century maps by Johann Baptist Homann and Jacques-Nicolas Bellin, the supplementation probably took place during that period. The map of Antwerp and environs shown here is by Claes Janszoon Visscher, the head of another prominent Amsterdam family of mapmakers who flourished at the same time as the Blaeus.

Koeman, 1967, vol. III, pp. 150, 176 [4].

32. Joan Blaeu, 1596–1673

Atlas Major, das ist, Welt-Beschreibung. Volume 8 of 9, *Hispannien . . . Africa, und America*

Amsterdam: 1667

ITEM 32. $16^{3}/_{8}$" X 21"

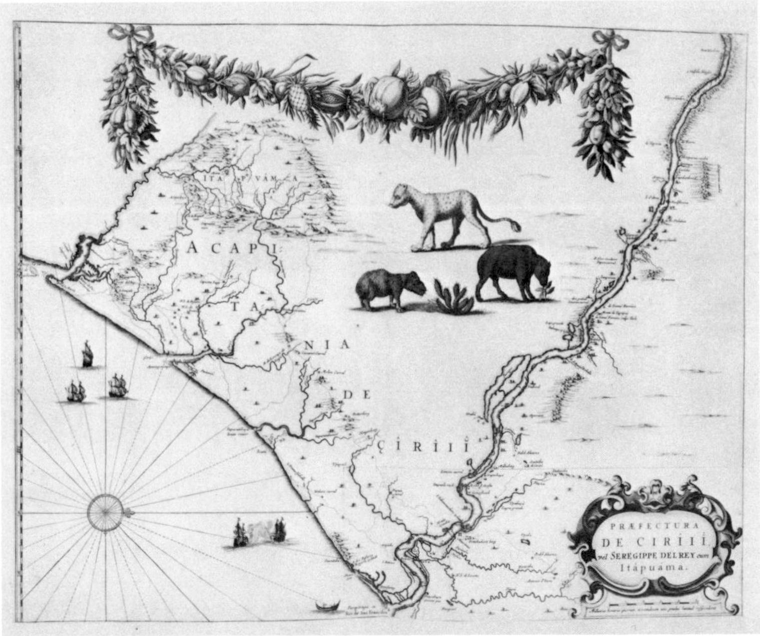

This regional map (*Praefectura de Cirîiî, vel Seregippe del Rey cum Itâpuáma*) is one of a set showing the former Dutch colonies in Brazil that was originally prepared by Blaeu for Caspar Barlaeus's *Rerum per Octennium in Brasilia* (1647). The decoration is from designs by the painter Franz Janszoon Post (ca. 1612–1680).

Koeman, 1967, vol. I, Bl 56 [616].

33. Joan Blaeu, 1596–1673

Atlas Major, das ist, Welt-Beschreibung. Volume 9 of 9, *Asia und Sina*

Amsterdam: 1667

Part of the success of the Blaeu family was the state patronage that Willem Janszoon Blaeu, and then his sons Joan and Cornelis (fl. 1638–1645), cultivated and acquired. Willem Janszoon was appointed hydrographer to the V.O.C. (the Dutch United East India Company) in 1633, and upon his death in 1638, Joan succeeded to this office. This map of the Spice Islands in the East Indies reflects the special Dutch interest in the area.

Koeman, 1967, vol. I, Bl 1 [59].

34. Joan Blaeu, 1596–1673 and Cornelis Blaeu, fl. 1638–1645

Territorii Francofurtensis Tabula

Amsterdam, 1667

From Joan Blaeu, *Atlas Major*, 1667

The Blaeu family of mapmakers (Willem Janszoon, Joan, and Cornelis) may not always have had the best and latest geographical information and may not have been willing or permitted to use what they had, but they gave their customers every conceivable substitute in the form of fine engraving, decorative cartouches, heraldic emblems, and magnificent coloring. This map of Frankfurt am Main and environs, first issued in 1640, is a dramatic example of such enhancement.

Koeman, 1967, vol. I, Bl 31A [293].

35. Frederik De Wit, 1630–1706

Nova Orbis Tabula, in Lucem Edita

Amsterdam: ca. 1680

De Wit was another accomplished Amsterdam engraver and map seller who flourished in the final quarter of the seventeenth century. His world map is one of the most attractive of its time. The corner vignettes combine images of the four seasons, the elements, and signs of the zodiac. The style of coloring (*en grisaille*) is very unusual for maps of this period—shades of grey were very seldom used—but the effect is striking and beautiful.

Koeman, 1967, vol. III, Wit 2 (1); Shirley, no. 499.

36. Carel Allard, ca. 1647–1706

Totius NeoBelgii Nova et Accuratissima Tabula

Amsterdam: ca. 1690

This map of what is now the northeast United States was first published circa 1650 by Jan Jansson and then copied many times over the course of the next century. Its prototype was representative of Dutch elegance and was also the most up-to-date map of the region available. The same could be said for the copy executed (ca. 1655) by Nicolaes Visscher, who was the first to add the view of New York. However, by the time that Huych Allard, Carel's father, first produced this version of the map, which provides a new *Restitutio* view of New York marking the Dutch recapture of the city in 1673, the image had lost some of its elegance and given way to the clutter that would become a characteristic of the bold and full engraving style favored by Germans in the eighteenth century.

Campbell, 1980; Koeman, 1967, vol. I, pp. 31–32.

37. John Speed, 1552–1629

The Theatre of the Empire of Great-Britain . . . Together with a Prospect of the Most Famous Parts of the World. New Edition

London: Printed for Thomas Basset and Richard Chiswel, 1676

2 parts in 1 volume

This exemplar of British mapmaking is displayed with Dutch cartography because all of its plates were engraved in Amsterdam. The plates for the *Theatre* were designed and engraved either by, or under the supervision of, Jodocus Hondius between 1605 and 1610. Those for the *Prospect* were also engraved in Amsterdam, and most bear the signature of Abraham Goos. Speed's *Theatre*, first published in 1611, contains British county maps, while the *Prospect*, first published in 1627, covers the rest of the world. Extremely popular, the books went through many editions. This example is the last complete edition; however, many of the plates would continue to be used well into the eighteenth century. It is a measure of Holland's superiority as a center of mapmaking that a London publisher would commission an Amsterdam engraver to produce his plates.

Skelton, 1970, pp. 30–36; *World Encompassed*, no. 171; Phillips, no. 5928.

American Type Founders Company Library

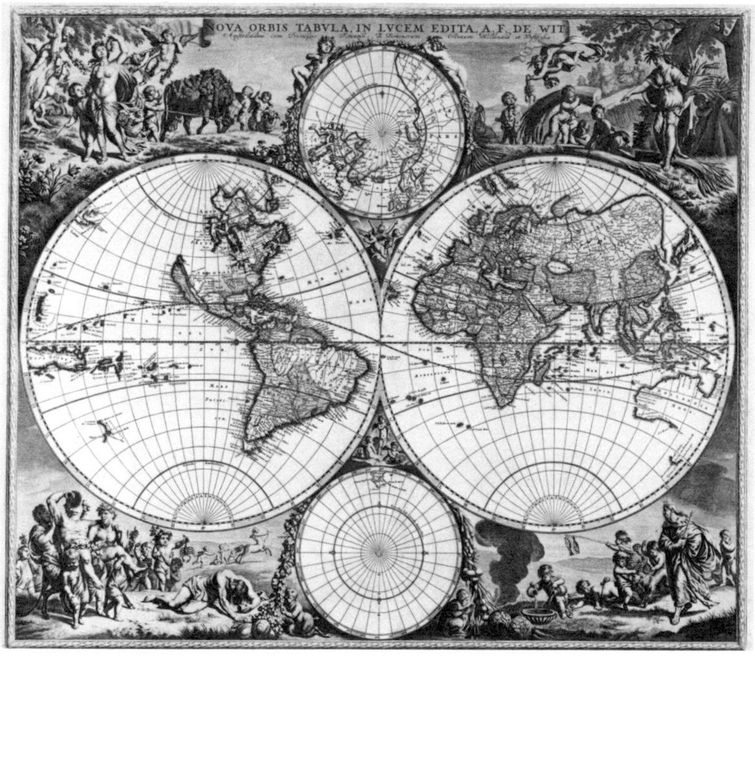

ITEM 35. 18¾" X 22¼"

ITEM 40. 19 7/16" X 25 3/4"

V. THE EMERGENCE OF SCIENTIFIC CARTOGRAPHY IN FRANCE

38. Nicolas Sanson, 1600–1667

L'Amérique en plusieurs cartes, & en divers traittés de géographie, et d'histoire

Paris: Chez l'autheur, 1657

France would supersede the Netherlands as the center of mapmaking in the eighteenth century, led by the Cassini family, Guillaume Delisle, and Robert de Vaugondy. This great age of French cartography may be said to have originated with Nicolas Sanson in the seventeenth century. The establishment of royal authority and patronage under Louis XIV created conditions under which the fine and applied arts could thrive. While the beginnings of the "French school" were not a wholly indigenous development—Sanson and all his engravers were either Flemings or from the neighboring province of Picardy—Sanson's maps have a distinctive appearance and a "wholly French elegance." This atlas of America was one of four continental volumes published by Sanson and his son. California is depicted as an island, a misconception that persisted throughout the seventeenth century.

Pastoureau, pp. 387–389, Sanson II A; Skelton, 1952, p. 67; Fordham, p. 24; *World Encompassed*, no. 157.

Bancroft fund

39. Nicolas de Fer, 1646–1720

L'atlas curieux

Paris: Chez l'auteur, 1700–1705

Nicolas de Fer would become one of the most productive mapmakers of his generation, with an output of over one thousand separate maps, plans, and views. For all that, he had a late start to his career, which did not begin in earnest until he was forty-one, when his widowed mother finally deeded him her late husband's modest print and map shop. Soon after taking over the business, de Fer acquired the patronage of the *grand Dauphin*, and he prospered immediately. *L'atlas curieux* was issued serially in six parts between 1700 and 1705. The map shown, *Les costes aux environs de la Rivière de Misisipi*, was issued in 1700. It records recent discoveries that provided a better representation of the Mississippi delta.

Pastoureau, pp. 167–170, Fer I.

Bancroft fund

40. Jean-Baptiste Nolin, 1648–1708

Carte des cinq provinces de l'Assistance de France des RR PP de la Compagnie de Jésus

Paris: J. B. Nolin, 1706

This map shows the administrative divisions of the Jesuits in France at the beginning of the eighteenth century. It records for each location the year in which a Jesuit presence was first established (beginning with the first *voeux* at Monmartre in 1534), the kind of presence (e.g. seminary, professed house, college), and even, through a set of codes, what was taught at each (e.g., G=grammar, T=theology). Based on the work of François de la Chaize and Claude François Menestrier, it was engraved by Jean-Baptiste Nolin and first published in 1706. (The map is signed by J. B. Nolin, "*le fils*"—suggesting a later date of issue for this copy.)

The senior Nolin was a highly skilled engraver who ultimately failed as a geographer and cartographer. Although appointed *géographe du roi* in 1701, he had little talent as a cartographer and resorted to plagiarism in constructing his maps. In 1706 he was convicted of plagiarizing Guillaume Delisle's work; he died in disgrace two years later.

Pastoureau, p. 357; Lagarde, pp. 21–25; Dainville.

41. Jean Picard, 1620–1682

Mésure de la terre

In Picard, *Dégré du méridien entre Paris et Amiens*

Paris: G. Martin, J. B. Coignard & H. L. Guerin, 1740

The origin of scientific cartography in France coincides with the founding of the Académie Royale des Sciences by Colbert in 1666. The main purpose of the Académie was to correct and improve maps and sea charts. An important step toward this goal was the accurate measurement of the earth's circumference and the calculation of the linear equivalent of a degree of arc. Jean Picard, a founding member of the Académie, was given this job. His method was to survey, by triangulation, a north-south line between two terminal points (Paris and Amiens) and then measure the arc (the difference in latitude) between them through astronomical observation. The method proved to be very accurate, and the results provided a standard value for a degree of longitude that would be used in subsequent work of the Académie. The present volume gives a detailed account of Picard's work.

Brown, pp. 208–218.

42. Jacques Cassini, 1677–1756

Planisphère terrestre

Paris: Jean-Baptiste Nolin, 1696

The measurement of longitude (distance east or west of a prime meridian) was a preoccupation of the scientific community in the late seventeenth century. No real progress in navigation or mapping could be made without a solution to this difficult problem. While a professor of astronomy at the University of Bologna, Jean-Dominique Cassini devised a practical method for measuring longitude (originally suggested by Galileo) that involved observations of Jupiter and its moons. It was after a paper on this method by Cassini came to the attention of the Académie that he was lured to Paris. The Académie obtained, by celestial observation, coordinates of latitude and longitude for locations all over the earth. As they collected the data, the results were transferred onto the floor of the Paris Observatory, where a giant planisphere twenty-four feet in diameter, with the North Pole in the center and meridians radiating out from the pole, had been inscribed. The map here, drawn by Cassini's son Jacques and engraved by Nolin in 1696, is a much reduced copy of the Paris Observatory "living map." Asterisks show the locations that had by then been fixed by observation. Only a handful of copies of this engraved map are known to exist. [BACK COVER ILLUSTRATION]

Brown, pp. 216–224; *World Encompassed*, no. 159; Shirley, no. 579.

43. Jean Picard, 1620–1682

Autograph letter, signed, to Jean-Dominique Cassini

Brest, 2 October 1679

Picard and Cassini collaborated closely with one another at the Académie Royale des Sciences. Here Picard reports from Brest, near the westernmost point in France, his observations on the timing of the eclipse (*immersion*) of Jupiter's first and second satellites, the basis of Cassini's method for determining longitude. Although the theory was well developed, the method had only recently been put into practice. In reference to the observations made on the second satellite, Picard remarks: "If the weather was as favorable to you as it was to us, we will be able to see what one can do for longitudes with this satellite."

Brown, pp. 218–224.

Gift of David Eugene Smith

44. César François Cassini de Thury, 1714–1784, and J. P. Maraldi

Nouvelle carte qui comprend les principaux triangles qui servent de fondement à la description géométrique de la France

Paris: 1744

In Cassini de Thury, *Description géométrique de la France*

Paris: Desaint, 1783

Colbert was anxious to construct an accurate map of France—the absence of which, he felt, was a national disgrace. A survey of the country was thus on the Académie Royale's agenda from the very start (1666). However, because of the enormous scale of the project and occasional lapses in state support, the detailed topographical survey would not be completed for more than a century. Four generations of Cassinis would eventually be involved in its execution. Both the elder Cassini and his son Jacques (1677–1756) were involved in extending Picard's meridian northward and southward by triangulation. Later, beginning in 1733, Jacques and his son, César François (1714–1784), began the eight-year survey that would produce a framework of great triangles that covered the whole of France. This resulted in the map shown here—the first complete topographic outline of any country based on triangulation and astronomical observations. The fourth generation, Jacques Dominique Cassini (1748–1845), would see the completion of the detailed field survey in 1789 and its eventual publication.

World Encompassed, no. 162; Brown, pp. 241–257; Konvitz, pp. 1–31.

Gift of Samuel Dale

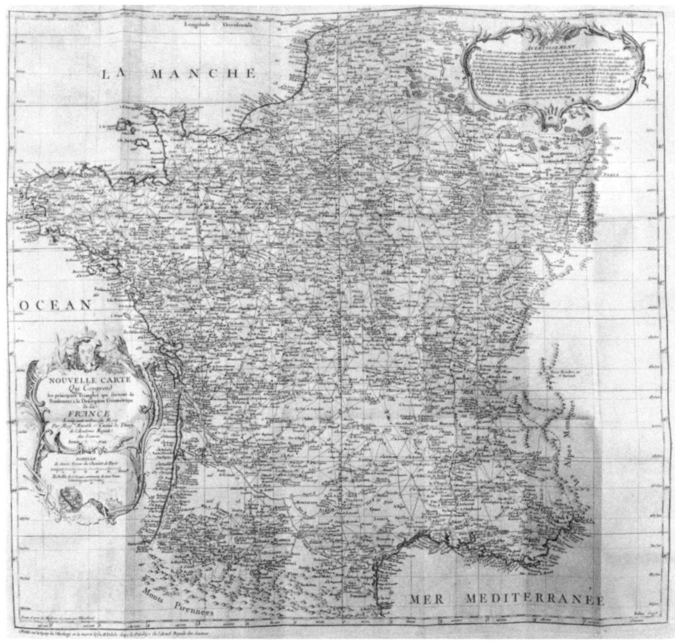

ITEM 44. 22 3/8" x 24 3/4"

45. César François Cassini de Thury, 1714–1784

France

Paris: Picquet, 1803

Volume 18 of 30

Although fieldwork for the detailed topographical survey of France was completed by 1789, the resulting maps would not be available to the public for twenty-five years. In 1793, when all but sixteen sheets were finished, the project was expropriated by the Dépôt Général de la Guerre. Security concerns and bureaucratic inertia restricted circulation of the maps to the military and government until the end of the Napoleonic Wars. The complete work is an impressive achievement. The scale is more than one inch to the mile, and if the sheets were joined together the whole would measure over one thousand square feet.

Konvitz, pp. 22–31.

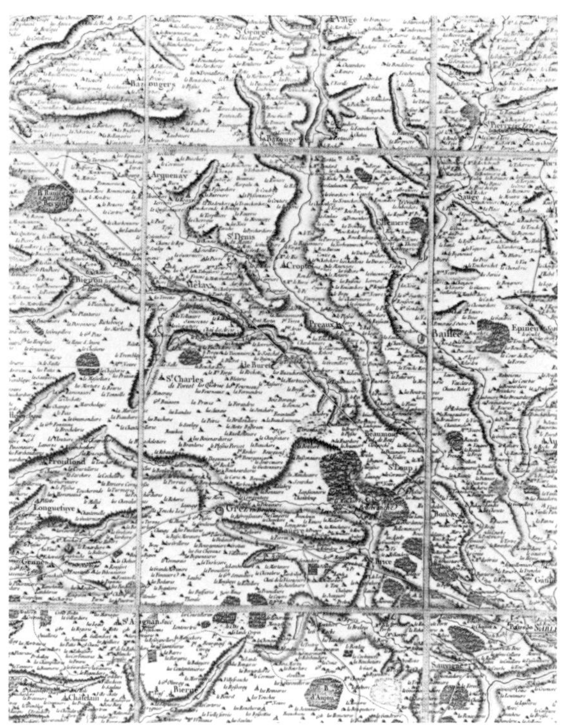

ITEM 45. 7 7/8" x 6 1/8" CENTER PANEL

46. Jacques Dominique Cassini, 1748–1845

Autograph letter, signed, to Nicholas-Antoine Sanson

Paris, 19 June 1803

ITEM 47. $17^3/_4$" DIAMETER

ITEM 48. 4" x $4^{13}/_{16}$"

When the army first expropriated Cassini's map in 1793, the government had no intention of compensating him or the other shareholders (of what had become the Compagnie de la Carte Générale de la France) for their loss. After receiving numerous petitions, however, the government relented and agreed to indemnify the company at a rate of 9,060 livres per share. But this sum was paid to only four of fifty shareholders before a new law, in December 1797, summarily reduced the value of a share to 166 livres. There matters rested until February 1803, when Napoleon acknowledged, by decree, the state's debt to the shareholders and ordered it paid.

In this letter, Cassini suggests to General Sanson (who, as director of the Dépôt Général de la Guerre, had physical possession of the map) that Napoleon's decree implied more than the payment—that it required the return of the plates and maps to the original subscribers. His interpretation was not shared by the government, and the Carte Générale de la France remained with the army. Cassini did win a moral victory of sorts when, in 1818, the Council of Ministers acknowledged his and the other shareholders' rights and awarded a further cash payment to compensate their loss.

Konvitz, p. 30.

Gift of David Eugene Smith

47. Guillaume Delisle, 1675–1726

Hémisphère Septentrional pour voir plus distinctement les terres arctiques

Amsterdam: R. & L. Ottens, 1740

Guillaume Delisle was a geographical child prodigy blessed with both a critical mind and an early exposure to the teaching of Jean-Dominique Cassini and the scientific culture of the Académie Royale. Deeply skeptical of received wisdom and traditional practice, he lost little time removing hoary myths, like California as an island, from his maps. He was one of the first to show a sustained tolerance for blank space in unknown parts of the world. By the age of twenty-five he had produced maps of the world and four continents consistent with observations made at the Paris Observatory; these established completely new standards for geography and cartography. It was Delisle who was most responsible for elevating France's position in European mapmaking. His work was frequently imitated in other countries. The example shown is a Dutch reissue of a map originally produced by Delisle in 1714.

Brown, pp. 242–245; Tooley, 1980, pp. 4–6.

48. Giovanni Antonio Rizzi-Zannoni, 1736–1814

Atlas géographique

Paris: Lattré, 1762

Although Italian, Rizzi-Zannoni was hydrographer to the Dépôt de Marine in Paris, and his early output, including this exquisite little atlas, was published in Paris. He later moved to Naples. Rizzi-Zannoni is widely admired for his fine engraving.

Moreland and Bannister, pp. 72–73; Tooley, 1978, p. 21.

Bancroft fund

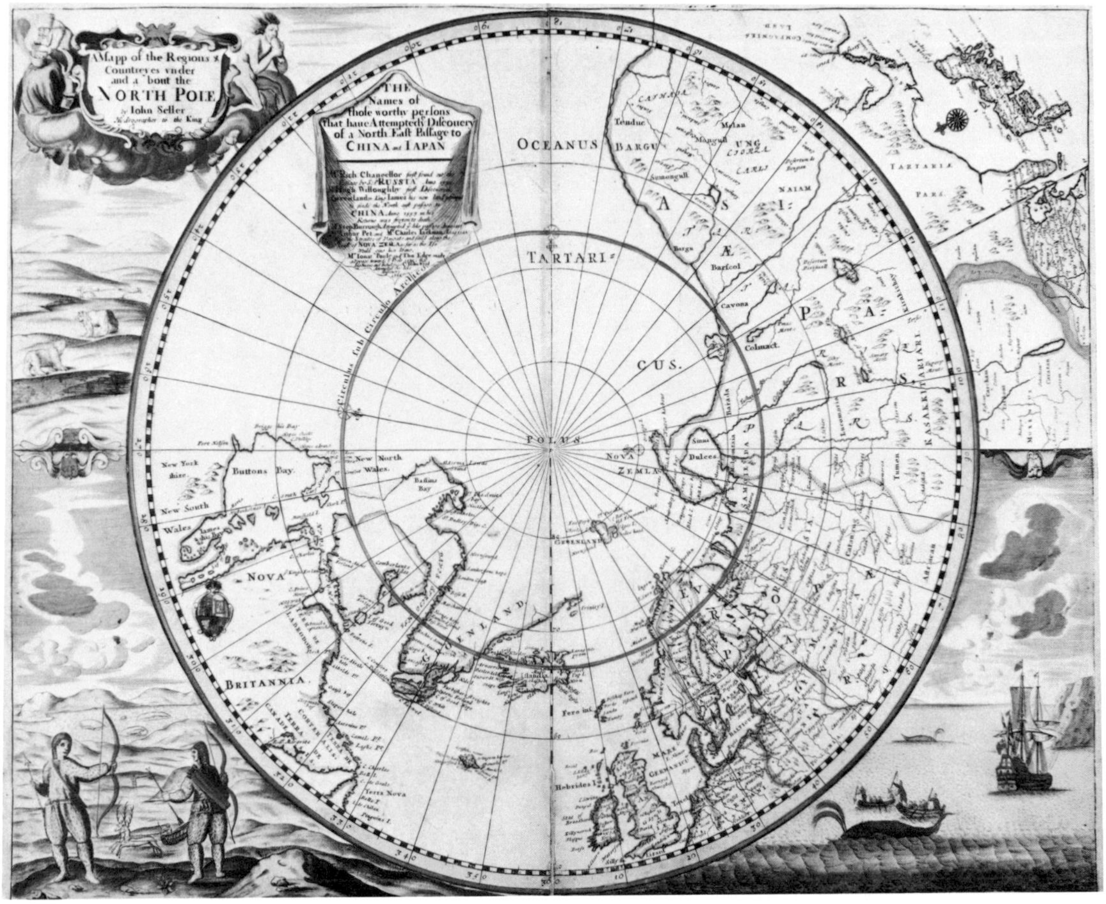

ITEM 49. $17^{1}/_{8}$" x $21^{3}/_{8}$"

VI. LONDON MAP SELLERS

49. John Seller, fl. 1658–1698

A Mapp of the Regions and Countreyes Under and About the North Pole

London: 1676

London had an active map trade in the last half of the seventeenth century, but the quality of most of the maps produced there did not compare favorably with continental work. The London mapmakers tended to be more tradesmen than scholars, and there was nothing like the state patronage of cartography that existed in France at this time.

John Seller, maker of instruments and maps, is credited with initiating the sea-chart trade in London with his *English Pilot*. First published in 1671, with charts advertised as "New and Exact," the *Pilot* used Dutch copperplates made in 1620 (by Jan Jansson), which in turn were copies of 1608 plates. *The English Pilot* was steadily expanded and updated and eventually passed into the more capable hands of John Thornton. Republished throughout the eighteenth century, much of Seller's original obsolete material continued to be mixed in with the new. The map of the North Pole shown here was published in his *Atlas Maritimus*.

Phillips, no. 4154; Tyacke, pp. 139–140; Verner; Howse and Sanderson, pp. 65, 67; *World Encompassed*, no. 192.

50. John Seller, fl. 1658–1698

America

London: ca. 1680

Seller had over one hundred maps of this small size which he published together as a world atlas (*Atlas Terrestris*) and in subsets by continent, like the present example. Atlases were often made to order at this time.

Phillips, no. 529; Verner, p. 148.

51. John Speed, 1552–1629

England, Wales, Scotland, and Ireland Described and Abridged

London: George Humble, 1627

Bound with:

A Prospect of the Most Famous Parts of the World

London: Printed by M. F. for W. Humble, 1646

Like Ortelius, John Speed published miniature versions of his large atlases. Speed's English county and world atlases were often published together, as in this volume, and are known as the "pocket Speed." Most of the county maps are from plates by Pieter van den Keere of Holland and are dated 1599 or later.

Skelton, 1970, no. 17.

52. Richard Blome, d. 1705

Britannia

London: Printed by T. Roycroft for R. Blome, 1673

Blome's popular *Britannia* was both a geography and heraldry of Great Britain. The county maps are attractive, if insubstantial. The *Dictionary of National Biography* says of *Britannia*, "Truly described by Bishop Nicolson as a 'most entire piece of theft out of Camden & Speed.'"

World Encompassed, no. 170; Skelton, 1970, pp. 139–145, 247.

53. Robert Morden, d. 1703

Geography Rectified: or, A Description of the World

London: Printed for Robert Morden and Thomas Cockerill, 1688

The cost of compiling a folio-size atlas was beyond the resources of London's Restoration publishers. In this context, and considering the amount of recycling of obsolete maps that was occurring at the time, Morden's humble *Geography Rectified* should be admired as one of the few new works of the period. It supplied separate maps of several areas, including Japan, not usually accorded such treatment. This is the second edition.

Campbell, 1972, no. 167; Phillips, no. 498.

Bancroft fund

54. John Senex, fl. 1702–1740

A Plan of the City's of London, Westminster, and Borough of Southwark

London: 1720

This handsome map of London was first published by John Senex in his popular *New General Atlas* in 1721. Along with his rival Herman Moll, Senex dominated the London map trade in the first quarter of the eighteenth century.

Darlington and Howego, no. 65.

55. Herman Moll, fl. 1678–1732

Atlas Minor

London: Printed for Thos. Bowles & John Bowles, 1745

Herman Moll, who migrated to England in about 1680 from his native Germany, was a far more original cartographer than his contemporaries in London. In a long career, which began as an engraver, he had an enormous output, including this *Atlas Minor*, first published in 1727, which proved particularly popular. Although biographical details for Moll are scarce, much can be inferred about his personality from the often acerbic notes and observations that he scattered across his large-scale maps. He was peculiarly hostile to the work of Guillaume Delisle and was, therefore, often wrong. A firm believer that California was an island, Moll wrote indignantly in 1711, "Why, I have had in my office mariners who have sailed around it."

Tooley, 1980, p. 111; Moreland and Bannister, pp. 163–164; Campbell, 1981, p. 37; Reinhartz.

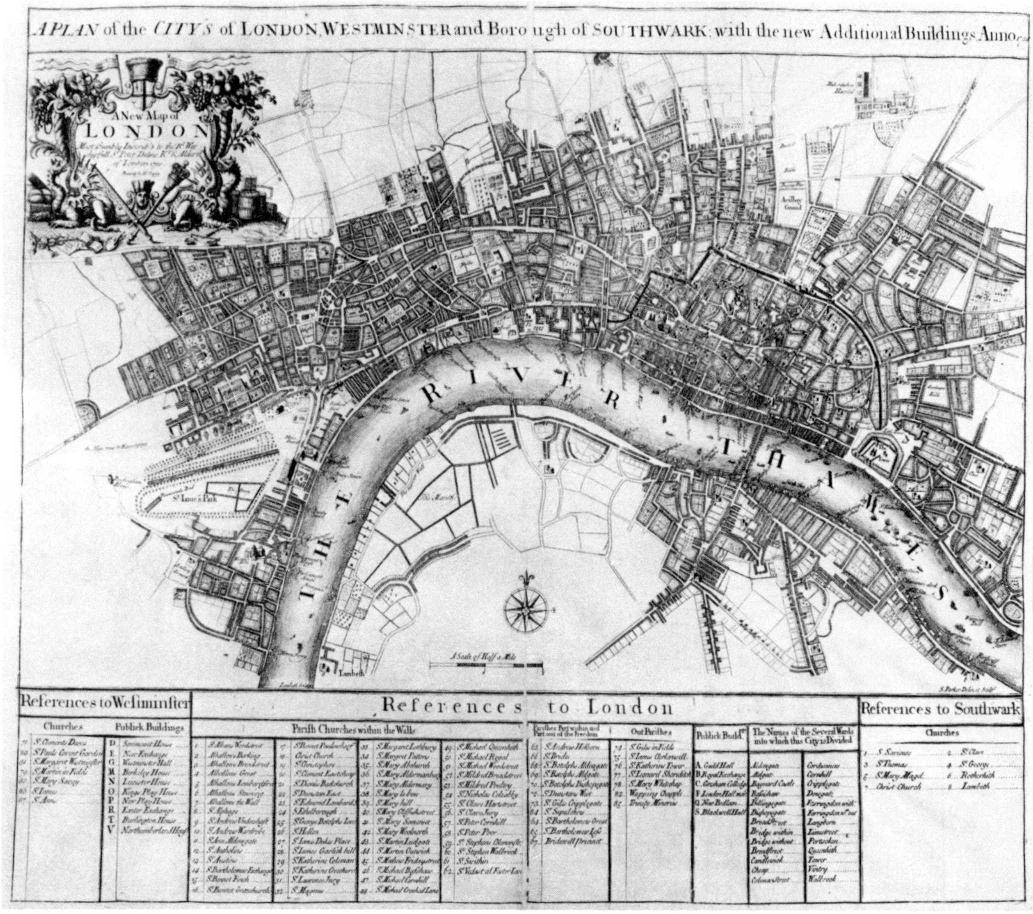

ITEM 54. 19 5/8" X 22 7/8"

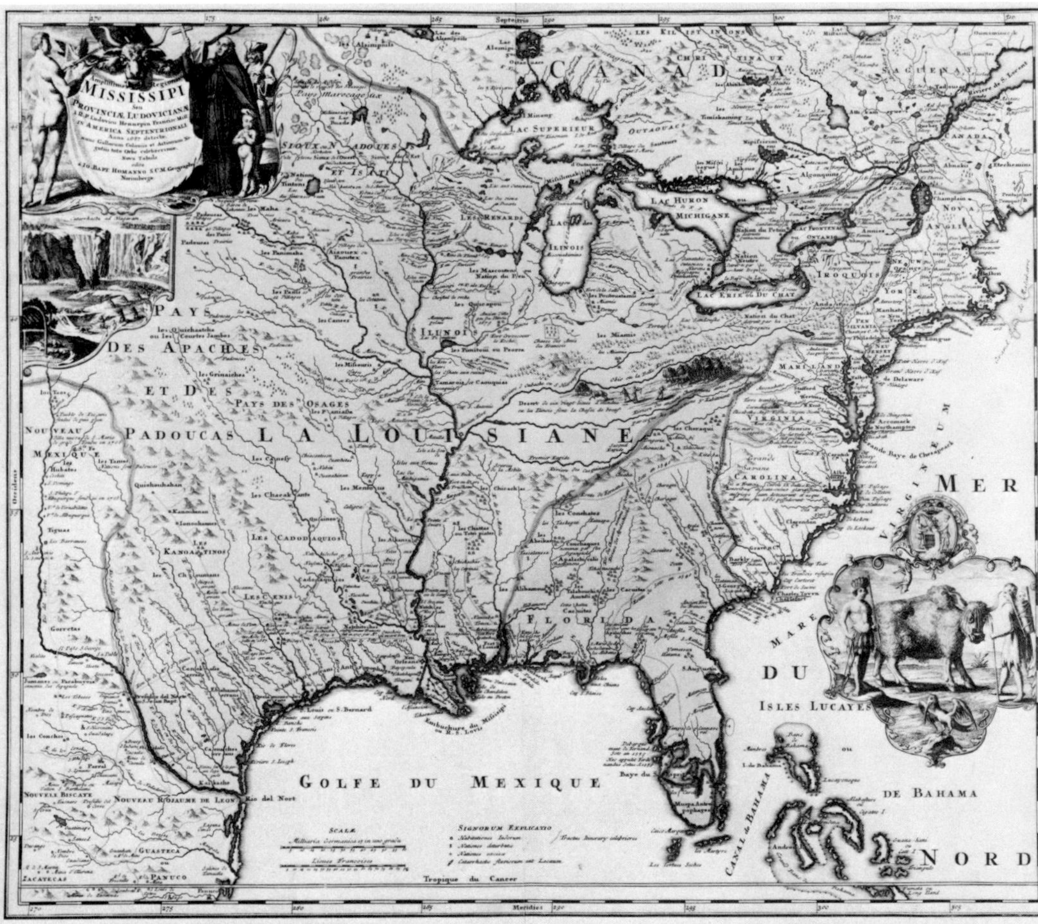

ITEM 56. 19 1/8" x 22 13/16"

VII. THE MAP TRADE IN EIGHTEENTH-CENTURY GERMANY

56. Johann Baptist Homann, 1663–1724

Amplissimae Regionis Mississippi seu Provinciae Ludovicianae a R. P. Ludovico Hennepin . . . Nova Tabula

Nuremberg: J. B. Homann, ca. 1724

Germany, even more than France and Italy, looked to Amsterdam for its maps in the seventeenth century. The resurgence of a strong indigenous German cartography began with Johann Baptist Homann, who established a printing house specializing in maps in Nuremberg in 1702. Homann achieved great commercial success by producing maps at a lower cost than the French and Dutch, from whom he copied liberally and judiciously. He was appointed geographer to the emperor in 1715. With his successors the enterprise became a center of geographical scholarship, attracting the astronomer J. G. Doppelmayer (1677–1750), who helped the firm keep abreast of the latest developments. This map is based largely on Guillaume Delisle's seminal *Carte de la Louisiane et du cours de Mississipi* published in 1718.

Skelton, 1952, p. 62; Crone, p. 83; Moreland and Bannister, pp. 84–86.

57. Johann Baptist Homann, 1663–1724

Regni Mexicani seu Novae Hispaniae, Ludovicianae, Novae Angliae, Carolinae, Virginiae et Pennsylvaniae . . . Accurata Tabula

Nuremberg: ca. 1724

Homann again copies Guillaume Delisle's work—in this case his *Carte du Mexique et de la Floride*, first published in 1703. The present example follows a revised edition of that map published by Jean Covens and Corneille Mortier in 1722.

Reinhartz, p. 30.

58. Georg Matthäus Seutter, 1678–1756

Atlas Novus

Augsburg: 1725

Seutter served an apprenticeship with Homann before establishing his own successful business in Augsburg in 1707. His maps display the Teutonic heaviness of Homann's style and are easily recognized by the distinctive way in which they are colored.

Phillips, no. 5968; Moreland and Bannister, p. 87; Blumer, p. 85.

59. Georg Matthäus Seutter, 1678–1756

Atlas Novus

Augsburg: ca. 1744

The representation of mountains and hilly terrain on maps has been a problem for cartographers from the

ITEM 59. 22½" x 19¼"

very beginning. The various solutions that have been proposed provide a fascinating study. For Europeans, the Alps have been the proving ground for the different approaches. Seutter's efforts in this map of the Swiss canton of Uri suggest how difficult the challenge is.

60. Tobias Conrad Lotter, 1717–1777

Topographia Sedis Imperatoriae Moscovitarum Petropolis

Augsburg: 1744

Tobias Lotter, a publisher and engraver, was Georg Matthäus Seutter's son-in-law. He worked closely with Seutter, engraving many of the latter's maps, and eventually succeeded to his father-in-law's business.

Moreland and Bannister, p. 88.

61. Tobias Conrad Lotter, 1717–1777

Atlas Minor

Augsburg: ca. 1744

This small atlas was issued simultaneously by both Seutter and Lotter, each using his own imprint.

Phillips, no. 3491.

Gift of George A. Plimpton

ITEM 62. 11 1/4" x 14"

VIII. THE SETTLEMENT OF NORTH AMERICA

62. Samuel Purchas, 1577?–1626

Purchas his Pilgrimes

London: Printed by W. Stansby for H. Fetherstone, 1625

Volume 3 of 4

Purchas's collection of travels and voyages includes Henry Briggs's *A Treatise of the North-west Passage to the South Sea*, which is illustrated by this attractively engraved map of North America. It is one of the first maps to show California as an island, and it is the first to show the name "Hudson Bay," although it is misplaced. There is no attempt yet to map the Mississippi or the Great Lakes. This copy of *Purchas* belonged to the English admiral Sir John Narbrough (1640–1688) and contains his annotation regarding his passage through the Strait of Magellan in 1670.

World Encompassed, no. 205; Schwartz and Ehrenberg, p. 101.

63. Sir William Vaughan, 1577–1641

The Golden Fleece

London: Printed for Francis Williams, 1626

3 volumes in 1

John Mason, who drew this, the first separate map of Newfoundland (*Insula Olim Vocata Nova Terra. The Iland Called of Olde Newfound Land*), was governor there in 1615. The Latin inscription summarizes important events in the island's history and notes the recipients of land patents, one of whom was Vaughan.

World Encompassed, no. 216; Schwartz and Ehrenberg, p. 99.

Seligman Library

64. Melchisedech Thevenot, 1620–1692

Carte de la découverte faite l'an 1673 dans l'Amérique Septentrionale

In Thevenot, *Recueil de voyages*

Paris: E. Michallet, 1681

This is the first published map to record the discoveries of Louis Joliet and Jacques Marquette in the upper Mississippi, Missouri, and Ohio rivers in 1673. It is also the first printed map of the Mississippi to be based on direct observation and the first to use the name "Michigan" in reference to one of the Great Lakes. The Mississippi is shown from the Gulf of Mexico to the Wisconsin River; Lake Michigan is to the east.

World Encompassed, no. 211; Schwartz and Ehrenberg, pp. 126–130.

Bancroft fund

65. Charles Mason, 1730–1787, and Jeremiah Dixon

A Plan of the Boundary Lines between the Province of Maryland and the Three Lower Counties on Delaware with Part of the Parallel of Latitude Which Is the Boundary between the Provinces of Maryland and Pennsylvania

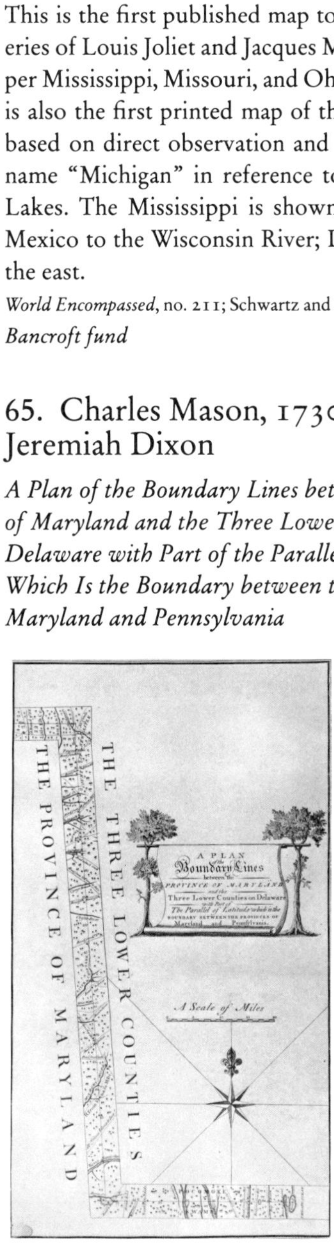

ITEM 65.
26 5/8" x 15 5/16"
ONE SHEET OF TWO

Philadelphia: Robert Kennedy, 1768

The boundary dispute between the Penns and Calverts, which began with the founding of Pennsylvania in 1682, simmered and flared for eighty years until it was finally resolved by the agreement that produced the famous Mason-Dixon line in 1768. Charles Mason, a British astronomer employed at the Greenwich Observatory, and Jeremiah Dixon, about whom very little is known, were appointed to survey a boundary line that was agreed to in 1763. Their progress westward along the arc of parallel that now forms the north-south boundary between Pennsylvania and Maryland was halted in 1767 by "Indian hostility," about twenty miles short of their goal (Dunkard's Creek). The line, which reached a length of two hundred and forty-four miles, was completed by others. The engravings were executed in Philadelphia in 1768 by Henry Dawkins and James Smithers. Whatever satisfaction was gained by the disputants through the agreement would be short-lived: the American Revolution would radically alter the ownership of the property in question.

Two hundred copies of the maps were printed for the Boundary Commission. Many of these were cut and backed so that a single continuous map of the boundary could be created. The present copy, which belonged to Benjamin Chew (1722–1810) of the Pennsylvania delegation to the Boundary Commission, is in its original, uncut form.

Lohf; Schwartz and Ehrenberg, pp. 168–170; *World Encompassed*, no. 254.

Gift of the Chew family through the courtesy of John T. Chew

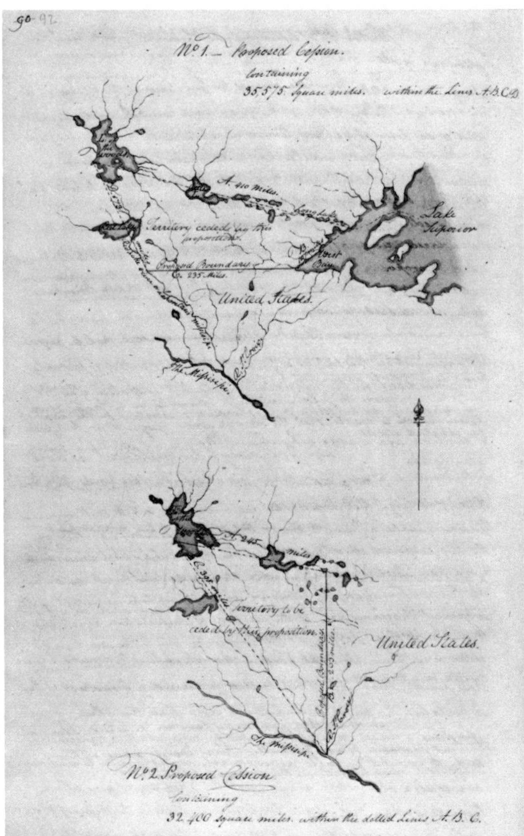

ITEM 66. 12½" x 8"

66. John Jay, 1745–1829, and John Trumbull, 1756–1843

Jay Treaty Letterbook

April 9, 1794–March 5, 1795

Manuscript letterbook with two ink and wash maps by John Trumbull

The celebrated painter John Trumbull accompanied John Jay to London in June 1794 to act as his secretary while he negotiated what would become the Treaty of Amity, Commerce, and Navigation, or the Jay Treaty. Fearing that an unintended gap existed in the 1783 northwest boundary between American and British territory, Lord Grenville, the British foreign secretary, wanted it redrawn. A gap would have prevented British access to the navigable part of the Mississippi, and, hence, to the Gulf coast. The maps by Trumbull sketch two of Grenville's proposals, both of which would have meant the loss of more than thirty thousand square miles of territory. Jay agreed to neither proposition and instead proposed a joint commission to settle the line if a gap really existed.

Morris and Servet; Fite and Freeman, p. 251.

Bancroft fund and various donors

IX. SEA CHARTS

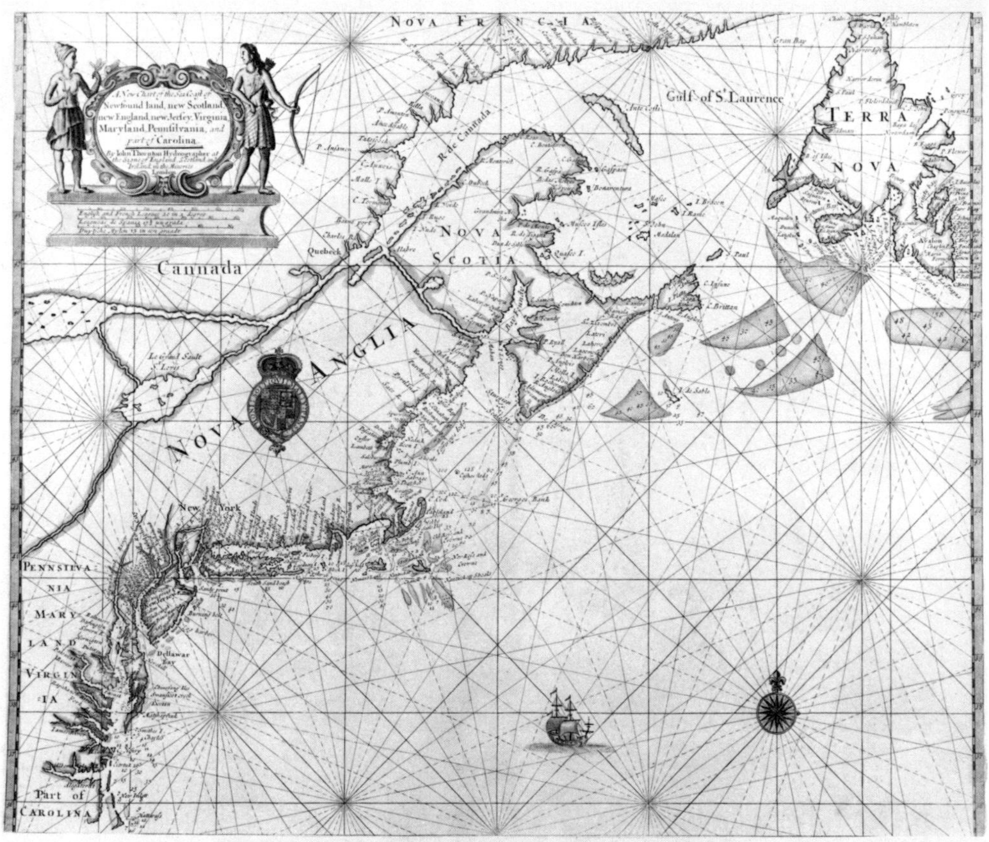

ITEM 69.
17" x 20 3/4"

67. Lucas Janszoon Waghenaer, 1533–1606

Der erst Theil dess Spiegels der Seefart . . . Auss niederlandischer in hochteutscher Sprach getrewlich ubersetzt . . . Durch Richart Slotboem

Amsterdam: Cornelius Claesz, 1589

The first publication of Waghenaer's *Spiegel der Zeevaert* in 1584 was a landmark in the history of the sea chart. It was the first printed sea atlas to assemble charts and sailing directions in a way that gave the mariner all that he needed to know about a particular area in one source. Among its many innovations were soundings (in fathoms at half-tide), the distortion of the coastline to allow more detail for such difficult passages as harbors and river mouths, and the placement of coastal views on the chart itself. Intended for piloting, the atlas covers the coasts of Europe from Norway to Spain. Lucas Janszoon Waghenaer was a pilot and hydrographer from Enkhuizen. The early editions of his *Spiegel* were printed in Leyden by Christopher Plantin. The present copy is the first German edition, printed for the first time in Amsterdam by Cornelius Claesz. The plates were somewhat worn by this time, and the impressions taken from them do not all show the magnificent engraving to its best effect.

Koeman, 1967, vol. IV, Wag 6, pp. 486–488; *World Encompassed*, nos. 181–183; Howse and Sanderson, p. 41; Koeman, 1964.

68. Eberhard Werner Happel, 1647–1690

Die Ebbe und Fluth auff einer flachen Landt-Karten fürgestelt

In Happel, *Relationes Curiosae*

Hamburg: Thomas von Wiering, 1683–1685

Volume 2 of 5

This unimposing little map, first published in 1675, has been identified as the earliest attempt to chart the main ocean currents of the world. Little is known about Happel or whether the map is the result of his own or another's scholarship. *Relationes Curiosae* is a massive, eclectic assemblage of fact and fancy, in the tradition of the *Nuremberg Chronicle*.

World Encompassed, no. 198.

69. John Thornton, 1641–1708

A New Chart of the Sea Coast of Newfoundland, New Scotland, New England, New Jersey, Virginia, Maryland, Pennsilvania and Part of Carolina

London: 1689

John Thornton was a member of the Worshipful Company of Drapers, a London livery company that spawned a group of manuscript chart makers, known as the Thames school, who carried on the trade for more than a century. Their charts were typically drawn on vellum and mounted on wooden boards, and they were meant for practical use at sea. Thornton was the only member of the school to also publish printed charts in any quantity. When John Seller's *English Pilot* began to flounder, Thornton took on the project. He produced the *Fourth Book* (*West India Navigation*) in 1689 and the *Third Book* (*Oriental Navigation*) in 1703. These, like the other "books" of the *English Pilot*, would go through many subsequent editions in the eighteenth century. The present chart, from the *Fourth Book*, is quite scarce.

Campbell, 1973; Tooley, 1978, pp. 60–62; Howse and Sanderson, p. 75; Tyacke, pp. 144–145.

70. Gerard van Keulen, 1678–1728

A Chart of the Sea Coasts of New Nederland, Virginia, New-England and Penn-silvania

Amsterdam: G. van Keulen, 1734

The van Keulen family of chart makers became the eighteenth-century equivalent of the Blaeus, rivaling them in scope and vitality of business. Founded by Johannes van Keulen in 1678, the house achieved its greatest prominence under his son Gerard, who took over the business in 1704. Gerard van Keulen was appointed hydrographer to the Dutch East India Company in 1714, and van Keulens would retain this position until the Company was dissolved. The two principal works of the house were the *Zee-Fakkel*, in six volumes, each corresponding to different parts of the world, and the *Zee-Atlas*, a single volume that would reach a maximum of 185 sheets. The present chart appeared in the 1734 edition of the *Zee-Fakkel*, Part IV. It is from a reworked plate originally made by Hendrick Doncker in 1688 and acquired by Johannes van Keulen in 1693 when he purchased Doncker's entire stock-in-trade.

Koeman, 1967, vol. IV, Keu 113B, p. 385 [321], pp. 276–279; Don 23 [105]; Howse and Sanderson, p. 71; Moreland and Bannister, p. 118; Suarez, no. 53.

71. Jacques-Nicolas Bellin, 1703–1772

Carte réduite des costes de la Louisiane et de Floride

From Bellin, *L'Hydrographie Françoise*

Paris: 1764

This attractive chart of the coasts between Galveston Bay and the Bahamas is by J. N. Bellin, *Ingénieur hydrographe de la Marine*. In contrast to the British Navy, which in the eighteenth century relied on private sources, France established a state hydrographic service—*Le Dépôt des Cartes et Plans de la Marine*—in 1720. Bellin joined the service in 1722 at the age of eighteen and remained there throughout his life. First published in 1756, *L'Hydrographie Françoise*, from which this chart is taken, is an atlas of large charts for waters other than those of France. The grid of perpendicular meridians and parallels tells us that the chart is drawn on the Mercator projection, which was in common use by this time.

Phillips, no. 590; Howse and Sanderson, p. 93.

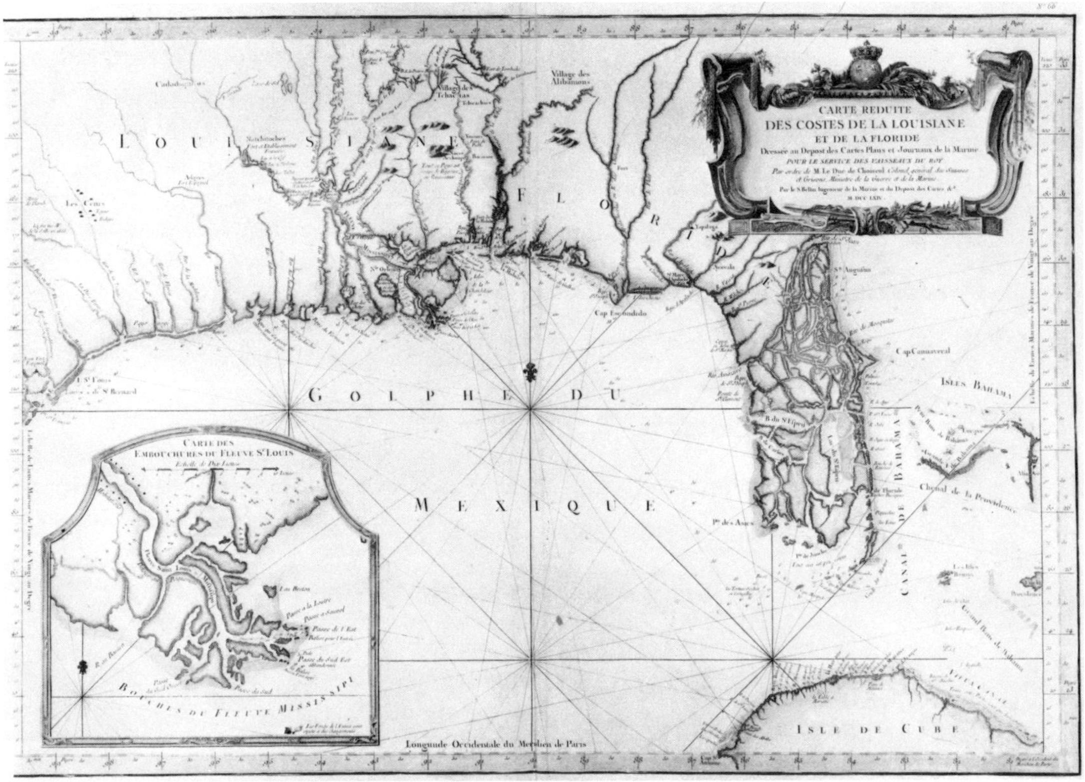

ITEM 71. 22 1/2" x 33 5/16"

72. Jacques-Nicolas Bellin, 1703–1772

Le petit atlas maritime

Paris: 1764

Volume 1 and 3 of 5

Another major work of Bellin was *Le petit atlas maritime*, with 580 smaller charts and harbor plans covering the whole world in five volumes. Bellin's small charts have a distinctive clean, spare, and elegant style, which belies the practical purpose for which they were intended.

Bancroft fund

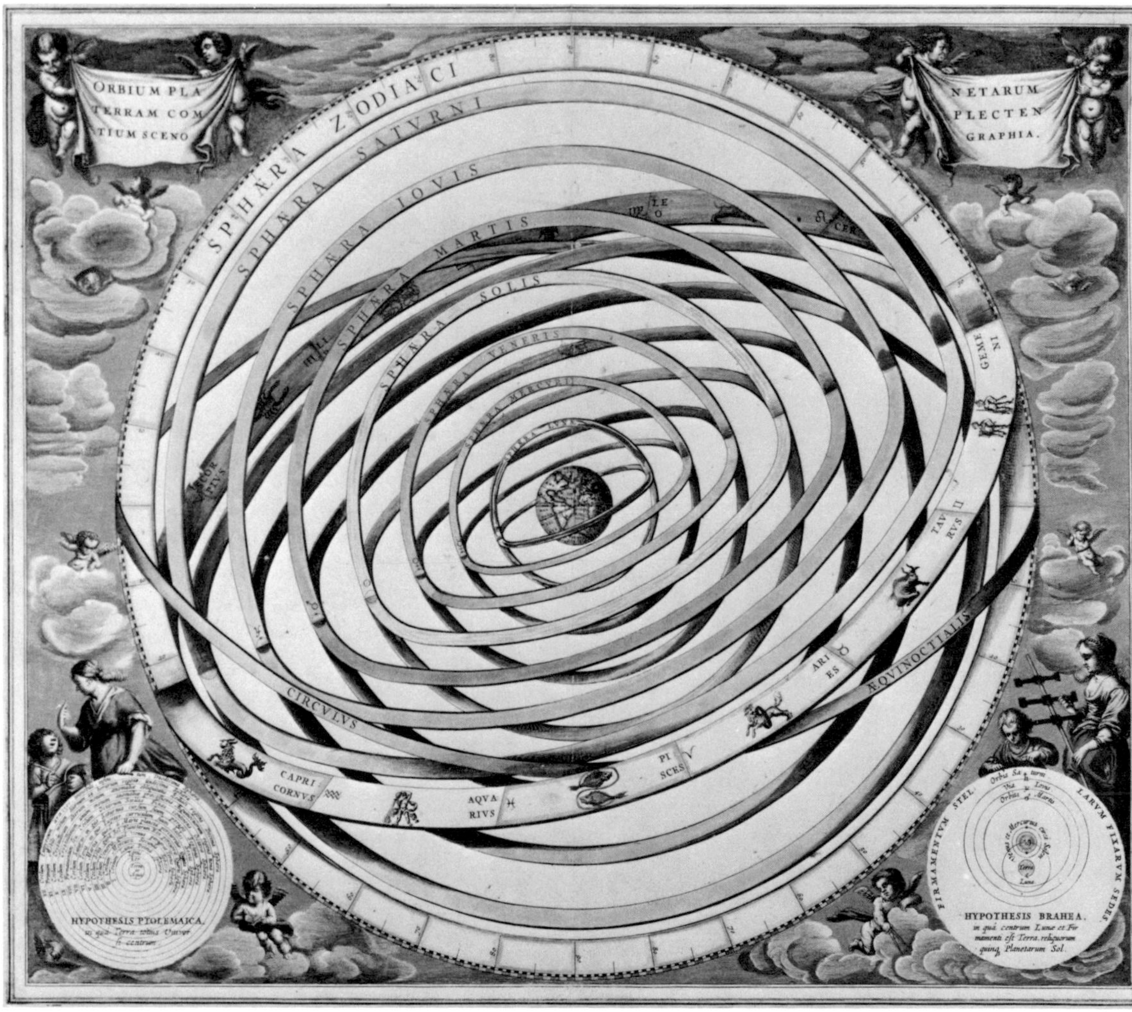

ITEM 75. $16^{7}/_{8}$" X $19^{15}/_{16}$"

X. CHARTING THE STARS

73. Andreas Cellarius fl. 1660

Scenographia Systematis Copernicani
From Cellarius, *Harmonia Macrocosmica*
Amsterdam: J. Jansson, 1660

Many would agree that the most beautiful images in this exhibit are the seventeenth-century celestial charts by Andreas Cellarius, combining as they do the finest Dutch design, engraving, and coloring. Little is known about Cellarius—only that he was rector of the Latin School at Hoorn in 1660 when the first edition of *Harmonia Macrocosmica* was published. No other celestial atlas was published in the Netherlands before the nineteenth century. The atlas contains twenty-nine charts that show the constellations and illustrate and compare the competing cosmological theories of Ptolemy, Copernicus, and Tycho Brahe; no text accompanies the charts. This chart shows the position of the earth relative to the sun at equinox and solstice.

Koeman, 1967, vol. IV, Cel 1, pp. 2–3.
Gift of Robert Halsband

74. Andreas Cellarius fl. 1660

Hemisphaerii Borealis Coeli et Terrae. Sphaerica Scenographia
From Cellarius, *Harmonia Macrocosmica*
Amsterdam: J. Jansson, 1660

The constellations of the northern sky are here superimposed on an image of the earth.
Gift of Robert Halsband

75. Andreas Cellarius fl. 1660

Orbium Planetarum Terram Complectentium Scenographia
From Cellarius, *Harmonia Macrocosmica*
Amsterdam: J. Jansson, 1660

This chart models the universe according to Ptolemy's conception. The two small spheres compare aspects of Ptolemy's universe with that of Tycho Brahe.
Gift of Robert Halsband

76. Johannes de Sacro Bosco, fl. 1230

Sphaera mundi
Paris: Wolfgang Hopyl, 1500

Sacro Bosco's famous text on astronomy and cosmography, *Sphaera mundi*, remained popular for four centuries because of its reliance on an established authority (Ptolemy's *Almagest*), its simplicity, and its compactness. Many manuscript copies have survived, and it went through more than sixty incunable editions after the first in 1472. The woodcut shows an armillary sphere—used to illustrate and analyze the various circles within the Ptolemaic celestial sphere.

Dilke, 1987, p. 188; Campbell, 1987, p. 105; *World Encompassed*, no. 13.

77. Peter Apian, 1495–1552

Cosmographia, sive Descriptio Universi Orbis, Petri Apiani & Gemmae Frisii
Antwerp: Jean Bellère, 1584

First published in 1524, Peter Apian's *Cosmographicus Liber* was an influential text and manual, which was kept in print throughout the sixteenth century. The purpose of cosmography was understood to be the description and mapping of the entire universe, and it thus subsumed astronomy and geography within its broad scope. The device shown here, Apian's *speculum cosmographicum*, first appeared in that earlier work. It consists of a circular world map (a polar stereographic projection) with attached volvelle and pointer and was intended for use in astrological calculations and for telling time. Apian developed many articulated devices such as this, as aids for explaining the Ptolemaic system, which would soon be swept aside by the Copernican revolution. The present volume is an edition of Apian's *Cosmographicus* by one of his students Gemma Frisius (1508–1555), first published in 1529.

Karrow, pp. 52, 62, 205.

Gift of David Eugene Smith

78. Vincenzo Maria Coronelli, 1650–1718

Epitome cosmografica
Colonia: 1693

Coronelli here provides a compact instruction manual for the use, elucidation, and construction of celestial spheres, terrestrial globes, astrolabes, and maps. Coronelli was himself a great map and globe maker, a successor to Giacomo Gastaldi as cosmographer to the Republic of Venice. He was justly famous for his globes, which varied in size from three inches in diameter to a pair of manuscript globes each fifteen feet in diameter. The latter, one celestial and the other terrestial, were made for Louis XIV in 1704.

World Encompassed, no. 127; Hough, no. 126.

BIBLIOGRAPHY

Blumer, Walter. *Bibliographie der Gesamtkarten der Schweiz von Anfang bis 1802*. Bern: Kommisionsverlag Kummerly and Frey, 1957.

Brown, Lloyd Arnold. *The Story of Maps*. Boston: Little, Brown, 1949.

Campbell, Tony. *Catalogue 9: Atlases*. London: Weinreb and Douwma, 1972.

———. "The Draper's Company and Its School of Seventeenth Century Chart-Makers." In Wallis and Tyacke, 1973, pp. 81–106.

———. "The Jansson-Visscher Maps of New England." Chapter 8 in Tooley, 1980, pp. 279–294.

———. *Early Maps*. New York: Abbeville Press, 1981.

———. *The Earliest Printed Maps, 1472–1500*. London: British Library, 1987.

Crone, G. R. *Maps and Their Makers: An Introduction to the History of Cartography*. 5th ed. Folkestone: W. Dawson, 1978.

Dainville, François de. "Les cartes anciennes de l'Assistance de France." *Archivum Historicum Societatis Jesu*, XIX, 1950, pp. 254–264.

Darlington, Ida and James Howego. *Printed Maps of London Circa 1553–1850*. London: George Philip and Son, 1964.

Dilke, O. A. W. *Greek and Roman Maps*. London: Thames and Hudson, 1985.

———. "Cartography in the Ancient World: A Conclusion." Chapter 16 in Harley and Woodward, 1987, pp. 276–279.

———. "Cartography in the Byzantine Empire." Chapter 15 in Harley and Woodward, 1987, pp. 258–275.

———. "The Culmination of Greek Cartography in Ptolemy." Chapter 11 in Harley and Woodward, 1987, pp. 177–200.

———. "Itineraries and Geographical Maps in the Early and Late Roman Empires." Chapter 14 in Harley and Woodward, 1987, pp. 234–257.

Fite, Emerson David and Archibald Freeman. *A Book Of Old Maps: Delineating American History from the Earliest Days Down to the Close of the Revolutionary War*. Cambridge: Harvard University Press, 1926.

Fordham, Sir Herbert George. *Some Notable Surveyors and Map-Makers of the Sixteenth, Seventeenth, and Eighteenth Centuries and Their Work: A Study in the History of Cartography*. Cambridge: The University Press, 1929.

Grafton, Anthony. *New Worlds, Ancient Texts: The Power of Tradition and the Shock of Discovery*. Cambridge, Mass.: Harvard University Press, Belknap Press, 1992.

Harley, J. B. *Maps and the Columbian Encounter: An Interpretive Guide to the Travelling Exhibition: American Geographical Society Collection, Milwaukee; James Ford Bell Library, Minneapolis; The Newberry Library, Chicago; William L. Clements Library, Ann Arbor*. Milwaukee: The Golda Meir Library, University of Wisconsin, 1990.

Harley, J. B. and David Woodward, editors. *The History of Cartography. Vol. 1. Cartography in Prehistoric, Ancient, and Medieval Europe and the Mediterranean*. Chicago: University of Chicago Press, 1987.

Harvey, P. D. A. "Local and Regional Cartography in Medieval Europe." Chapter 20 in Harley and Woodward, 1987, pp. 464–501.

Howse, Derek and Michael W. B. Sanderson. *The Sea Chart: An Historical Survey Based on the Collections in the National Maritime Museum*. Newton Abbot: David and Charles, 1973.

Hough, Samuel J. *The Italians and the Creation of America: An Exhibition at The John Carter Brown Library*. Providence: Brown University, 1980.

Karrow, Robert W. *Mapmakers of the Sixteenth Century and Their Maps: Bio-Bibliographies of the Cartographers of Abraham Ortelius, 1570*. Chicago: Published for The Newberry Library by Speculum Orbis, 1993.

Koeman, Cornelis. *The History of Lucas Janszoon Waghenaer and His "Spieghel der zeevaerdt."* New York: American Elsevier, 1964.

———. *Atlantes Neerlandici: Bibliography of Terrestrial, Maritime and Celestial Atlases and Pilot Books, Published in the Netherlands up to 1880*. 6 v. Amsterdam: Theatrum Orbis Terrarum, 1967–1985. (Vol. 6 published by Canaletto, Alphen aan den Rijn.)

Konvitz, Josef W. *Cartography in France, 1660–1848: Science, Engineering, and Statecraft*. Chicago: University of Chicago Press, 1987.

Lagarde, L. "Le Passage du Nord-Ouest et la Mer de l'Ouest dans la cartographie française du dix-huitième siècle: Contribution à l'étude de l'oeuvre des Delisle et Buache." *Imago Mundi*, 41, 1989, pp. 19–43.

Lohf, Kenneth A. "The Map That Marked The End of a Landmark American Boundary Dispute: A Gift from the Chew Family." *Columbia Library Columns*, XXXII, 2, 1983, pp. 29–31.

Moreland, Carl and David Bannister. *Antique Maps*. 2d ed. Oxford: Phaidon Christie's, 1986.

Morris, Richard B. and Ene Sirvet. "The Artist and Diplomat: John Trumbull and the Jay Mission to England." *Columbia Library Columns*, XXXI, 2, 1982, pp. 18–27.

Nordenskjöld, A. E. *Facsimile-Atlas to the Early History of Cartography; with Reproductions of the Most Important Maps Printed in the Fifteenth and Sixteenth Centuries.* Translated from the Swedish original by Johan Adolf Ekelof and Clements R. Markham. Stockholm: 1889; reprinted New York: Dover Publications, 1973.

Pastoureau, Mireille. *Les atlas français dix-huitième–dix-septième siècles: Repertoire bibliographique et étude.* Paris: Bibliothèque nationale, Departement des cartes et plans, 1984.

Phillips, P. L. and C. E. LeGear. *A List of Geographical Atlases in the Library of Congress, with Bibliographical Notes.* 9 vols. Washington: Govt. Print. Off., 1909–1992.

Reinhartz, Dennis. "Additions to the Gulf and Indies Maps of Herman Moll." *The Map Collector*, 43, 1988, pp. 27–30.

Rey Pastor, Julio and Ernesto García Camarero. *La cartographía mallorquina.* Madrid: Departamento de Historia y Filosofía de la Ciencia, "Instituto Luis Vives," Consejo Superior de Investigaciones Científicas, 1960.

Rücker, Elizabeth. "The Chronicle of the World Known As *The Nuremberg Chronicle* by Hartmann Schedel, Nuremberg 1493." *IMCOS Journal*, 55, 1993, pp. 8–19.

Ruland, Harold L. "A Survey of the Double-Page Maps in the Thirty-Five Editions of the *Cosmographia Universalis* 1544–1628 of Sebastian Münster and His Editions of Ptolemy's *Geographia* 1540–1552." *Imago Mundi*, 16, 1962, pp. 84–97.

Schwartz, Seymour I. and Ralph E. Ehrenberg. *The Mapping of America.* New York: Abrams, 1980.

Shirley, Rodney W. *The Mapping of the World: Early Printed World Maps 1472–1700.* London: Holland Press, 1983.

Skelton, R. A. *Decorative Printed Maps of the Fifteenth to the Eighteenth Centuries.* (A revised edition of *Old and Decorative Maps and Charts* by A. L. Humphreys.) London: Staples Press, 1952.

———. *County Atlases of the British Isles, 1579–1850: A Bibliography.* London: Carta Press, 1970.

———. *Maps: a Historical Survey of Their Study and Collecting.* Chicago: Published for the Hermon Dunlap Smith Center for the History of Cartography, the Newberry Library, by the University of Chicago Press, 1975.

Stevens, Henry Newton. *Ptolemy's "Geography": A Brief Account of All the Printed Editions down to 1730.* London, 1908; reprinted Amsterdam: Theatrum Orbis Terrarum, 1972.

Suarez, Thomas. *Shedding the Veil: Mapping the European Discovery of America and the World.* Singapore: World Scientific, 1992.

The World Encompassed: An Exhibition of the History of Maps Held at the Baltimore Museum of Art, October 7 to November 23, 1952. Baltimore: Trustees of the Walters Art Gallery, 1952.

Thrower, Norman J. W., editor. *The Compleat Plattmaker: Essays on Chart, Map, and Globe Making in England in the Seventeenth and Eighteenth Centuries.* Berkeley: University of California Press, 1978.

Tooley, R. V. *Maps and Map-Makers.* 6th ed. London: Batsford, 1978.

———, ed. *The Mapping of America.* London: Holland Press; New York: R. B. Arkway, 1980.

Tyacke, Sarah. *London Map-Sellers, 1660–1720: A Collection of Advertisements for Maps Placed in the London Gazette, 1668–1719, with Biographical Notes on the Map-Sellers.* Tring: Map Collector Publications, 1978.

Verner, Coolie. "John Seller and the Chart Trade in Seventeenth-Century England." Chapter IV in Thrower, 1978, pp. 127–157.

Wallis, Helen and Sarah Tyacke. *My Head Is a Map: Essays and Memoirs in Honour of R. V. Tooley.* London: Francis Edwards and Carta Press, 1973.

Woodward, David. "Reality, Symbolism, Time, and Space in Medieval World Maps." *Annals of the Association of American Geographers*, 75, 1985, pp. 510–521.

———. "Medieval Mappaemundi." Chapter 18 in Harley and Woodward, 1987, pp. 286–370.

Wolff, Hans. *America: Early Maps of the New World.* Munich: Prestel-Verlag, 1992.

INDEX OF NAMES

Numbers refer to items
Italic indicates main entries

750 COPIES OF THIS CATALOGUE
WERE PRINTED BY THE STINEHOUR PRESS IN
LUNENBURG, VERMONT. DESIGNED BY JERRY KELLY
EDITED BY RUDOLPH ELLENBOGEN AND CLAUDIA FUNKE
EXHIBITION DESIGN AND INSTALLATION BY
CLAUDIA FUNKE AND JANE SIEGEL
PHOTOGRAPHY BY MARTIN MESSIK

*This catalogue was made possible by
the generous support of Swann Galleries, Inc., Frank Streeter,
Herbert Robinson, Charles Tanenbaum, Martayan Lan Rare Maps,
and donors to the Rare Book and Manuscript Library Fund*